Modern Biotechnology

S.B. PRIMROSE BSc, PhD
Amersham International plc
Amersham, Buckinghamshire

Modern Biotechnology

Blackwell Scientific Publications
OXFORD LONDON EDINBURGH
BOSTON PALO ALTO MELBOURNE

Editorial offices:
Osney Mead, Oxford, OX2 0EL
8 John Street, London, WC1N 2ES
23 Ainslie Place, Edinburgh, EH3 6AJ
52 Beacon Street, Boston
Massachusetts 02108, USA
667 Lytton Avenue, Palo Alto
California 94301, USA
107 Barry Street, Carlton
Victoria 3053, Australia

First published 1987

DISTRIBUTORS

USA and Canada
Blackwell Scientific Publications Inc
P O Box 50009, Palo Alto
California 94303

Australia
Blackwell Scientific Publications
(Australia) Pty Ltd
107 Barry Street,
Carlton, Victoria 3053

Set by Setrite International Typesetters
Hong Kong
Printed and bound in Great Britain
at the University Press, Cambridge

British Library Cataloguing in Publication Data

Primrose, S.B.
Modern biotechnology.
1. Biotechnology
I. Title
660'.6 TP248.2

ISBN 0-632-01781-3
ISBN 0-632-01764-3 Pbk

Library of Congress Cataloging-in-Publication Data

Primrose, S.B.
Modern biotechnology.

Includes index.
1. Biotechnology. I. Title.
TP248.2.P75 1987 660'.6 86-32699
ISBN 0-632-01781-3
ISBN 0-632-01764-3 (pbk.)

Contents

Preface

Biotechnology is one of the buzz words of the decade. Rightly or wrongly it is often associated with easy money and, consequently, means all things to all people. So what is biotechnology? At its simplest it is the commercial exploitation of living organisms or their components, e.g. enzymes. Thus the science formerly known as industrial microbiology clearly falls within the definition of biotechnology and to many they are synonomous. But plants and animals are also exploited commercially in the practices known as horticulture and agriculture and these could come under the umbrella of biotechnology. By common consent the terms plant and animal biotechnology are restricted to that grey area encompassing the application of modern molecular biology and cell culture techniques to the manipulation of plants and animals, e.g. plant cell culture and transgenic animals, and which is not yet part of conventional plant and animal breeding.

This broad view of what constitutes biotechnology forms the basis of this book. The overriding aim has been to combine in a single text detailed information on recombinant DNA technology, protein engineering, industrial microbiology, monoclonal antibodies, plant and animal cell culture, new methods of plant and animal breeding and the legal, social and ethical issues which surround biotechnology. It provides a basic reference source for undergraduates taking courses in biochemistry, genetics and molecular biology and also for industrial specialists who seek further information on areas they are less familiar with.

The book is heavily biased towards the impact of in-vitro gene manipulation but then without gene manipulation biotechnology would not be a buzz word!

SANDY PRIMROSE

Acknowledgements

Few authors can produce the finished manuscript for a textbook entirely unaided and this author is no exception. Critical appraisal of various chapters was provided by numerous colleagues and Les Bell, Paul Hissey, Dave Lewis, George O'Neill and Steven Vranch deserve special mention. Omissions and errors highlighted by them have been corrected and they are absolved from any which remain. Paul Elias of Blackwell Scientific Publications and Dr Richard Walden of the University of Leicester read the entire draft text and their numerous comments were invaluable during the preparation of the final manuscript. Special mention is due to Maureen Bevan and Jill Redpath for the many (frustrating?) hours spent typing and correcting the numerous drafts of each chapter: hopefully they can appreciate this, the published text. Last, but most important of all, thanks are due to my wife and children for obligingly going on holiday on two occasions so that I could work intensively without interruption.

Abbreviations

AAT	α_1-antitrypsin
ADA	adenosine deaminase
ADH	alcohol dehydrogenase
BPV	bovine papilloma virus
CaMV	cauliflower mosaic virus
cDNA	copy DNA
DEAE	diethylaminoethyl
DHFR	dihydrofolate reductase
dpm	disintegrations per minute
ELISA	enzyme-linked immunosorbent assay
EPA	Environmental Protection Agency (USA)
EPSP	3-enolpyruvyl-shikimate 5-phosphate
FDA	Food and Drug Administration (USA)
FMDV	foot and mouth disease virus
GMP	good manufacturing practice
GVH disease	graft-versus-host disease
HAT medium	hypoxanthine, aminopterin and thymidine medium
HFCS	high-fructose corn syrup
HLA	human leucocyte antigens
HPLC	high-performance liquid chromatography
HPRT	hypoxanthine phosphoribosyl transferase
MAB	monoclonal antibody
MEOR	microbial-enhanced oil recovery
MMT	mouse metallothionein
MS medium	Murashige and Skoog medium
NIH	National Institutes of Health (USA)
NMR	nuclear magnetic resonance
rbs	ribosome binding site
RF	replicative form
RFLP	restriction fragment length polymorphisms
RIA	radioimmunoassay
RuBPCase	ribulose bisphosphate carboxylase
SCP	single-cell protein
Ti plasmid	tumour-inducing plasmid
TK	thymidine kinase
VSV	vesicular stomatitis virus
VVM	volume (of air) per volume (of liquid) per minute

Part I
Introduction

1/Biotechnology — Ancient and Modern

INTRODUCTION

Biotechnology is not a recent development. Microorganisms have been used to produce food such as beer, vinegar, yoghurt and cheese for over 8 millenia. Certainly the ancient Sumerians were familiar with beer and alehouses were an established part of Roman civilization. Wine also was popular with the Romans and they tried introducing grapevines into southern Britain for the express purpose of winemaking. References to wine and vinegar (Fig. 1.1) are scattered throughout the Bible, which is an indication that their production dates back to early times.

In many instances microbial contamination of food results in spoilage, although what is unpalatable to one race may be a delicacy to another! Occasionally microbial growth would result in beneficial changes such as improved flavour and texture and, more importantly, improved storage quality. Once these desirable changes had occurred, they would be self-perpetuating. In the absence of a knowledge of microbiology storage vessels would not be cleaned and the residual food would act as an inoculum. In many respects modern production of fermented foods is little different: open vessels are still used and a residue from one batch is used to inoculate the next one.

Ethanol was the first chemical to be produced with the aid of biotechnology. The origins of distillation are not clear but by the 14th century AD it was widely used to increase the alcoholic content of wines and beers. Indeed it was at this time when the 'auld alliance' between France and Scotland was at its prime that the French brandy manufacturers taught the Scots brewers to distil their beer to produce whisky! From the production of spirit beverages it was but a small step to the production of neat alcohol and approximately 25% of world ethanol production is still produced by this biological route.

20 And Noah began to be an husbandman, and he planted a vineyard:
21 And he drank of the wine, and was drunken; and he was uncovered within his tent.

Book of Genesis, Chapter 9.

46 And about the ninth hour Jesus cried with a loud voice, saying, Eli, Eli, lama sabachthani? that is to say, My God, my God, why hast thou forsaken me?
47 Some of them that stood there, when they heard that, said, This man calleth for Elias.
48 And straightway one of them ran, and took a spunge, and filled it with vinegar, and put in on a reed, and gave him to drink.

The crucifixion of Christ.
Book of Matthew, Chapter 27.

Fig. 1.1 Biblical references to alcohol and vinegar.

Until just over a century ago it was not realized that microorganisms were involved in the production of alcohol and vinegar. The discovery came when a group of French merchants were searching for a method that would prevent wine and beer from souring when they were shipped over long distances. They asked Louis Pasteur for help. At the time many scientists believed that air acted on the sugars in these fluids to convert them into alcohol. Instead, Pasteur found that yeasts convert the sugars to alcohol in the absence of air. Such an anaerobic process is known as *fermentation*. Souring and spoilage occur later and are due to the activities of a group of bacteria, the acetic acid bacteria, which convert alcohol into vinegar (acetic acid). Pasteur's solution was to heat the alcohol just enough to kill most of the microorganisms present, a process that does not greatly affect the flavour of the wine or beer. This process is known as *pasteurization*, although we now know that a similar technique was used for the manufacture of sake in the Orient over 300 years earlier.

THE FIRST WORLD WAR AND THE RISE OF THE MODERN FERMENTATION INDUSTRY

Distillation apart, there was little change in biotechnology from the pre-Christian era until the early 20th century and, as is often the case with technological advance, the impetus was provided by war. At the start of the First World War the British naval blockade prevented the Germans from importing the vegetable oils necessary to produce glycerol for explosives manufacture. Consequently the Germans turned to the microbial production of glycerol by yeast (Fig. 1.2) and soon were able to manufacture over 1000 tons per month. For its part, Germany was able to hinder the British war effort since prior to hostilities it had been the source of acetone and butanol: the former was required for munitions and the latter for artificial rubber. The result was the British development of the acetone–butanol fermentation (Fig. 1.3) using *Clostridium acetobutylicum*.

The glycerol fermentation was short-lived. By contrast, the acetone–butanol fermentation survived until the early 1950s and during the Second World War some interesting innovations were made, such as the introduction of semi-continuous fermentation. Initially *Cl. acetobutylicum* was cultured in a large volume of medium until growth ceased. At this stage two-thirds of the medium was removed for extraction of acetone and butanol and the culture vessel filled up with fresh medium. With

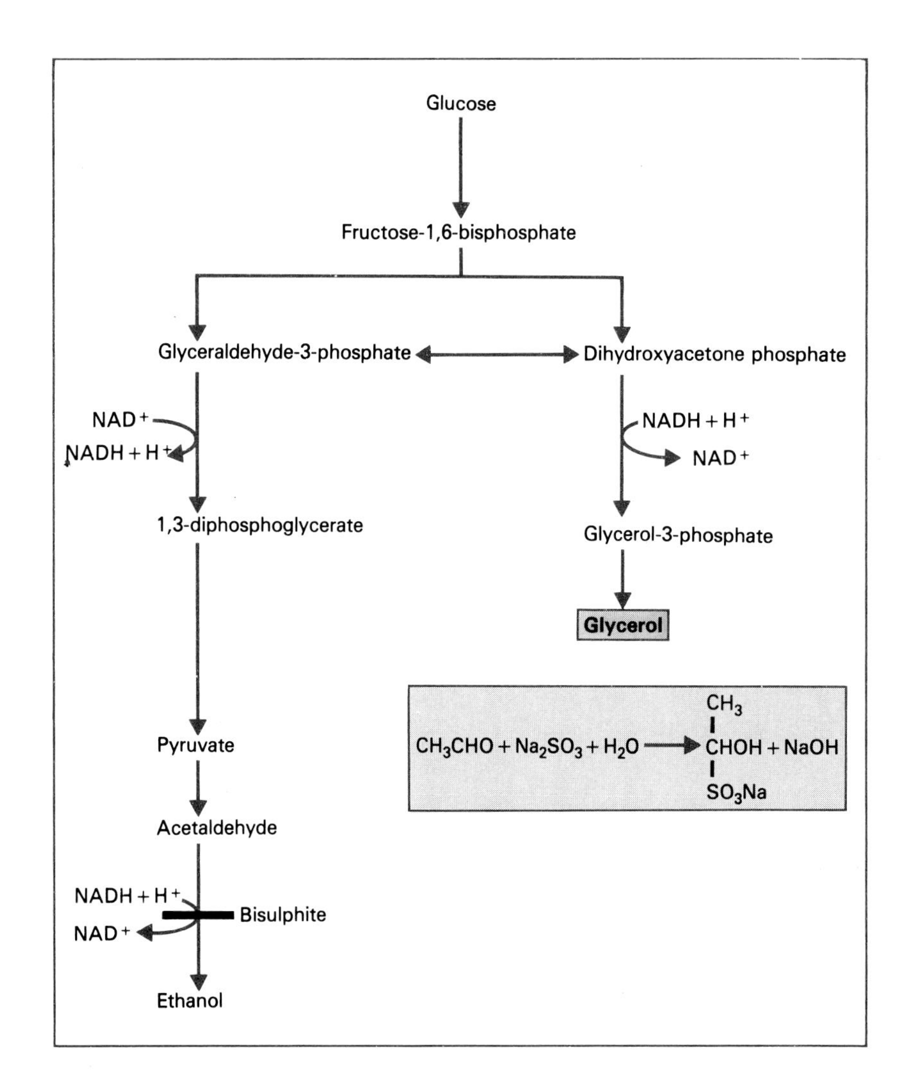

Fig. 1.2 The formation of glycerol instead of ethanol when bisulphite is added to a yeast culture. The bisulphite forms an addition product with acetaldehyde as shown in the inset.

such a large inoculum of cells complete conversion of sugar to solvents occurred in only 12 h. Later it was realized that it was possible to use non-sterile medium; although the yield of solvent was lower due to the growth of contaminating bacteria, the savings in fuel were substantial. In the immediate post-war period many organic chemicals, including acetone and butanol, became readily available from by-products of the petroleum industry and the fermentation process was discontinued. However, the acetone–butanol fermentation is an exceedingly simple process which might be of benefit to many Third World countries who cannot afford to spend vast sums of money on either petroleum itself or petrochemical-based products. Since these countries often have an abundance of the necessary cheap raw materials such as sugar or starch this fermentation process may make a comeback!

Present-day citric acid manufacture also has its origins in the First World War. Until then citric acid had been extracted from citrus fruits and the major producer was Italy. As men were called to arms the citrus groves were left untended. By the time hostilities ended the industry was in ruins and the price of citric acid had escalated. This paved the way for the introduction of a microbial process in 1923. Unlike *Cl. acetobutylicum*, the organism used to produce citric acid (*Aspergillus niger*) is an obligate aerobe and so must be cultured in the presence of oxygen. Initially large-scale culture was achieved by placing liquid medium in shallow metal pans and allowing the organism to grow on the surface. Later this method of *surface culture* was improved by absorbing the nutrient medium onto an inert granular support (Fig. 1.4(a)).

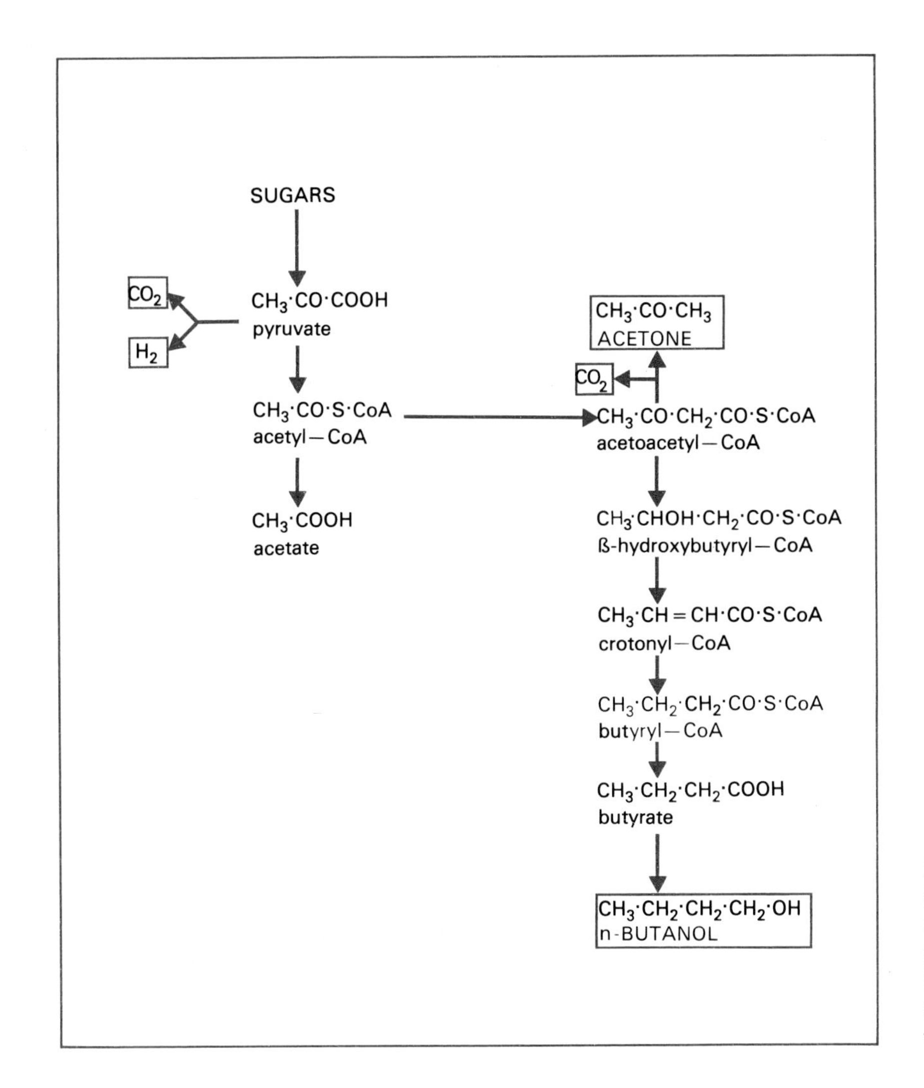

Fig. 1.3 The pathway of acetone and butanol formation by *Clostridium acetobutylicum*. Note that hydrogen gas is produced in addition to the two solvents. Not surprisingly, a number of manufacturing plants were destroyed by explosions!

(a)

(b)

Fig. 1.4 Production of citric acid (a) by surface culture and (b) in a stirred tank reactor. (Photos courtesy of John and E. Sturge Ltd.)

PENICILLIN AND THE PRODUCTION OF FINE CHEMICALS

Penicillin was the name given by Fleming to the antibacterial substance produced by the mould *Penicillium notatum*. By 1940 when the first purified preparations became available the phenomenal curative properties of penicillin were obvious. The significance of this discovery was not lost on a country again at war. Like the *A. niger* used to produce citric acid, *P. notatum* is an obligate aerobe and had to be grown in surface culture. Not only was this labour-intensive but the cultures were prone to contamination, thus reducing the yield of penicillin. The need for aseptic operation led to the development of the stirred tank reactor (Fig. 1.4(b)), which to this day is the preferred method for the large-scale cultivation of microbes. Aseptic operation is achieved by sterilizing all the equipment with steam prior to inoculation and by keeping the pressure inside the vessel higher than atmospheric pressure. To meet the oxygen demand of the culture, sterile air is blown into the vessel and distributed throughout the medium by agitation.

A second contribution to modern biotechnology that was made by the penicillin programme was the development of strain selection procedures. The original *P. notatum* culture yielded only 2 mg of penicillin per litre of culture fluid but by screening many different *Penicillium* isolates a higher-yielding variant, *P. chrysogenum*, was identified. In an attempt to improve the yield still further the *P. chrysogenum* was exposed systematically to a variety of mutagens such as nitrogen mustard, ultraviolet radiation and X-radiation. After each round of exposure the survivors were screened and the highest-yielding variant carried forward for the next round of mutations. By combining fermentation improvements with the use of mutants the titre has been increased to over 20 g/l.

That microbes could produce antibiotics had been known for a long time but was considered of little significance. Once the clinical utility of penicillin was established, pharmaceutical companies began to consider antibiotics seriously. Following the discovery that *Streptomyces griseus* also produces a clinically useful antibiotic, streptomycin, it became standard practice to screen large numbers of environmental isolates for their ability to produce antibiotics. Since then it has been shown that the filamentous bacteria known as actinomycetes, a group to which *S. griseus* belongs, produce many hundreds of different antibiotics, including at least 90% of those known today. Gradually the screens became more and more sophisticated and the need to test more and more microorganisms led to the isolation of microorganisms from increasingly exotic sources. In recent times it has become increasingly difficult to find antibiotics which are both novel and useful; the number of new antibiotics discovered per year has remained constant but over the last 20 years the clinical success rate has dropped from 5% to less than 1%. Consequently pharmaceutical companies have redirected their screening efforts towards the identification of pharmacologically active fermentation products. As with the penicillin fermentation, once a microbe is identified which produces a useful metabolite, strain improvement and fermentation development are undertaken.

INDUSTRIAL USES OF PLANT AND ANIMAL CELL CULTURE

Animal cell culture has long been used in the production of viral vaccines but until the 1960s large-scale culture had not even been attempted. It was successful due largely to the vigorous application of the principles of aseptic operation formulated during the development of the penicillin fermentation. With some animal cell lines there is a problem not encountered with microbial cells, that is that the cells do not grow in suspension but require a surface for growth. To satisfy this requirement some novel solutions have been adopted. In some instances the surface area inside the vessel is increased by the addition of microcarrier beads which are maintained in suspension; in other cases, multiple plates are fitted inside the fermenter. Despite these successes the application of mass animal cell culture has been limited largely to the production of vaccines. New avenues have opened up with the realization that certain cell lines derived from human tumours can secrete sufficiently high levels of human proteins to warrant their use for commercial production. Thus the Bowes melanoma line overproduces tissue plasminogen activator which can be used to dissolve blood clots following coronary thrombosis. The Namalwa cell line overproduces interferons and currently is a commercial source of these antiviral and anticancer proteins. Nevertheless, the number of tumour lines identified that produce useful products is limited.

In the last ten years certain tumour cell lines have attracted a great deal of attention and these are the *hybridomas*. As their name implies, hybridomas are hybrid cells. They are created by fusing myeloma (a type of tumour) cells with antibody-producing spleen lymphocytes. Following fusion with the myeloma cell the lymphocyte acquires immortality and can be grown indefinitely in cell culture while continuing to secrete antibody. Since any given lymphocyte only synthesizes a single antibody species, all the antibody molecules made by culture of any particular hybridoma will be identical. Since all the antibodies in the preparation are identical they are said to be *monoclonal*, i.e. they are all derived from a single clone of lymphocytes. Monoclonal antibodies can be purified easily and find application in many different areas from diagnostic kits to cancer therapy and protein purification.

For a long time the ability to cultivate plant cells in the laboratory was considered no more than a curiosity. Although whole plants serve as valuable sources of agricultural chemicals, drugs, flavourings and colourings, very few of these compounds are produced in cell culture. When they are so produced, yields are low and classical mutation and selection techniques have not been particularly successful in improving them. Today plant cell culture is of considerable importance and this is a direct result of the development of methods for the regeneration of plants from individual cells. The benefits of this are twofold. First, regeneration enables hundreds of plants to be produced in a single experiment. During regeneration genetic variants are thrown off at a high frequency and these can be very useful to the plant breeder. Second, virus-free cells can be isolated and used to produce virus-free crops which will give increased yields per acre.

GENETIC ENGINEERING AND THE NEW BIOTECHNOLOGY

After the introduction of the penicillin fermentation there were virtually no significant new developments in industrial microbiology for 30 years. Most companies followed the practice of screening microbes for desirable metabolites or activities and then, however low the levels, implementing strain selection and fermentation development. In the late 1960s considerable excitement was generated by the prospect of using microbial cells (or *biomass*) as a source of protein, the so-called *single-cell protein* or SCP. The rationale for this was that globally there was a shortage of protein and with a rapidly expanding population this situation was going to become worse. However the introduction of SCP has not been a success. Development coincided with a rapid increase in oil prices and the introduction of improved, high-yielding varieties of crops (Fig. 1.5). The developed countries did not need SCP, they had a plentiful supply of protein from conventional sources; the underdeveloped countries, on the other hand, could not afford to buy SCP or even to build and run SCP plants. Thus developments in biotechnology, as in other areas, are subject to political and economic pressures. High quality science does not guarantee commercial success and there is little altruism in the business world.

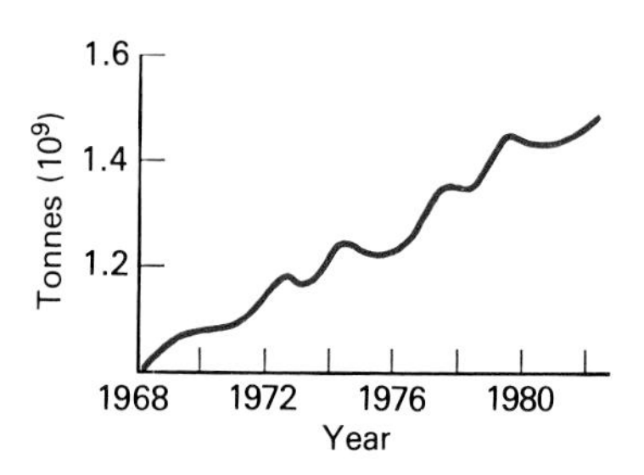

Fig. 1.5 Increase in grain production in the period 1968–1982. Similar increases in yield have been reported for other protein sources.

In the 1980s biotechnology has become a major growth area. This change has come about through a single development: the ability to splice together *in vitro* DNA molecules derived from different sources. This gene splicing ability is referred to as gene manipulation and because one is recombining pre-existing genetic sequences to create a novel combination the term *recombinant DNA technology* is also used. Whereas mutation and selection can be used to increase the level of pre-existing activity in a microbial cell, recombinant DNA technology can be used to confer on cells entirely *new* synthetic capabilities; for example, bacteria can for the first time synthesize human hormones. This technology is not confined to microorganisms: plant and animal cells and even intact plants and animals can also be modified.

Early work on the commercial applications of recombinant DNA technology centred on the production of proteins. There were two reasons for this. Not only did the technology enable many proteins, principally of therapeutic interest, to be produced for the first time but proteins are the immediate products of a gene. As genetic engineers have become skilled there has been a trend towards cloning all the genes associated with a biosynthetic pathway; for example, all the genes encoding the 29 steps in erythromycin biosynthesis have been cloned on a single DNA fragment. One outcome of this may be that ultimately the range of different organisms used in commercial biotechnology may diminish. Where low molecular weight compounds are required the pathways may be introduced to a limited selection of bacteria and fungi, e.g. *Escherichia coli*, *Streptomyces* sp., *Saccharomyces cerevisiae* and *Aspergillus* sp. Large molecular weight compounds may be produced in an even more restricted range of organisms and animal cells may grow in importance because they can effect post-translational modifications of proteins in an identical fashion to the intact animal (see p. 16).

Biotechnology, as practised today, is much more than recombinant DNA technology. It includes hybridoma technology, the use of cells and enzymes immobilized on inert supports and the ability to regenerate plants from isolated cells. However, if these are the performers, gene manipulation is the star of the show. The ability to splice genes has revolutionized the industry and is leading to the development of countless new products and improved methods for well-established processes. The aim of this book is to put into context all the new technologies and therefore no apology is made for the heavy emphasis on recombinant DNA technology. Having said that, it must be admitted that at the time of writing no company has made much money from the use of recombinant microorganisms but that monoclonal antibodies have lived up to their commercial promise.

Further reading

GENERAL

A general introduction to biotechnology can be found in the September 1981 issue of *Scientific American*, and the book by Steve Prentis *Biotechnology: a new industrial revolution* (1984, Orbis Publishing, London). The former concentrates on industrial microbiology, whereas the latter covers a similar range of topics to this book.

An up to date review of biotechnology but without the gene cloning emphasis is provided by I.J. Higgins, D.J. Best and J. Jones in their book *Biotechnology, Principles and Applications* (1985, Blackwell Scientific Publications, Oxford).

More advanced coverage is provided by the three special issues of *Science*, the official journal of the American Association for the Advancement of Science: volume 196 number 4286 concentrated on the methodology of gene manipulation, volume 209 number 4463 on the research applications of the techniques and volume 219 number 4585 on commercial biotechnology. There are a number of journals devoted solely to biotechnology. The best of these are *Bio/technology* (Macmillan, London) and *Trends in Biotechnology* (Elsevier, Amsterdam).

SPECIFIC

Hastings J.J.H. (1971) Development of the fermentation industries in Great Britain. *Developments in Applied Microbiology* **16**, 1–45.

Maddox J. (1986) New technology of medicine. *Nature* **321**, 807.

Part II
Recombinant DNA Technology

2/The Basic Principles of Recombinant DNA Technology

INTRODUCTION TO CLONING

Let us suppose that we wish to construct a bacterium that produces human insulin. Naïvely it might be thought that all that is required is to introduce the human insulin gene into its new host. In fact, a foreign gene is not maintained in its new environment if it is simply inserted into the bacterium on a DNA fragment for such fragments are not replicated (Fig. 2.1). The reason for this is that the enzyme DNA polymerase which makes copies of the DNA does not initiate the process at random but at selected sites known as *origins of replication*. Invariably, fragments of DNA do not possess an origin of replication. Using recombinant DNA technology it is possible to ensure the replication of the insulin gene in its new host by inserting the gene into a *cloning vector*. A cloning vector is simply a DNA molecule possessing an origin of replication and which can replicate in the host cell of choice. Most cloning vectors used with microorganisms are extrachromosomal, autonomously replicating circles of DNA called *plasmids* (Fig. 2.1). Viruses are used occasionally as vehicles for gene insertion into microorganisms (see Chapter 4) but they are far more important for work with animal cells (see Chapter 9).

In order to insert foreign DNA into a plasmid, use is made of special enzymes known as *restriction endonucleases*. These enzymes cut large DNA molecules into shorter fragments by cleavage at specific nucleotide sequences called *recognition sites* (Fig. 2.2), i.e. restriction endonucleases are highly specific deoxyribonucleases (*DNases*). Some of these enzymes cut the two helices a few base pairs apart generating two fragments with single-strand protrusions called *sticky ends* because their bases are complementary. Fragments of the foreign DNA are inserted into plasmid vectors cut open with the same enzyme or one which produces a matching end (Fig. 2.3). The resulting recombinants or *chimaeras* are introduced into the host cell by the process of transformation. In this process the DNA is mixed with a suspension of bacteria, which have been prepared under specialized conditions, and the DNA enters the cell by a mechanism (*transformation*) which still is poorly understood.

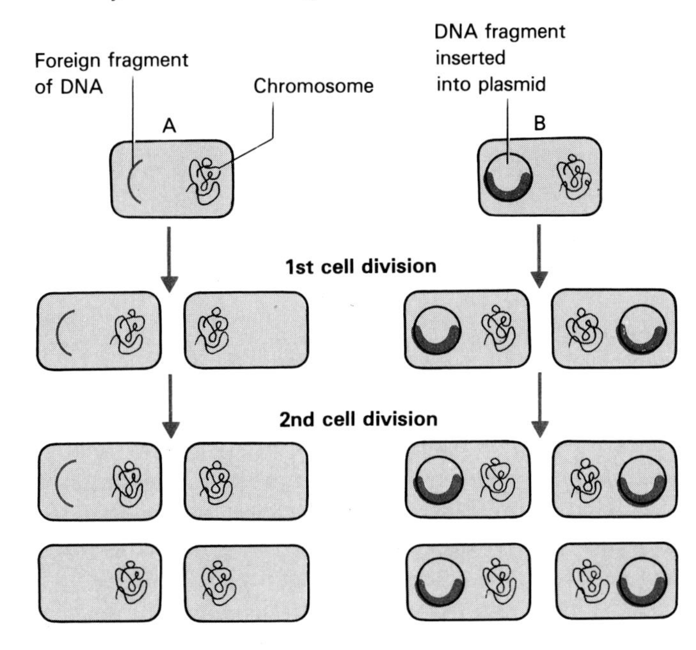

Fig. 2.1 The requirements for a cloning vector. (A) Fragments of DNA introduced into the bacterium by transformation do not undergo replication and are gradually diluted out of the population. (B) DNA fragments introduced into plasmids are inherited by both daughter progeny at cell divison.

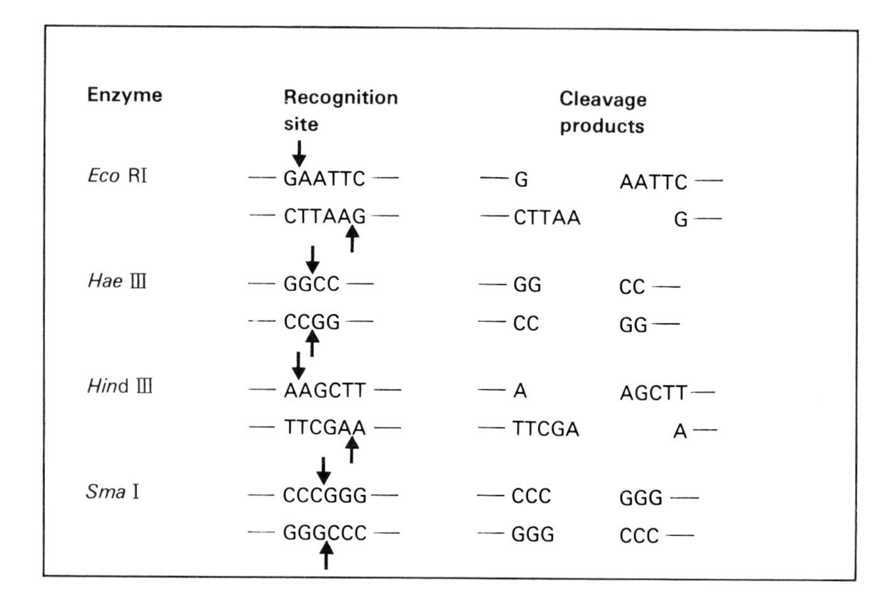

Fig. 2.2 The recognition sites for some common restriction endonucleases. The arrows indicate the cleavage points. Note that the enzymes *Sma* I and *Hae* III generate blunt ends, while *E. coli* and *Hind* III leave sticky ends.

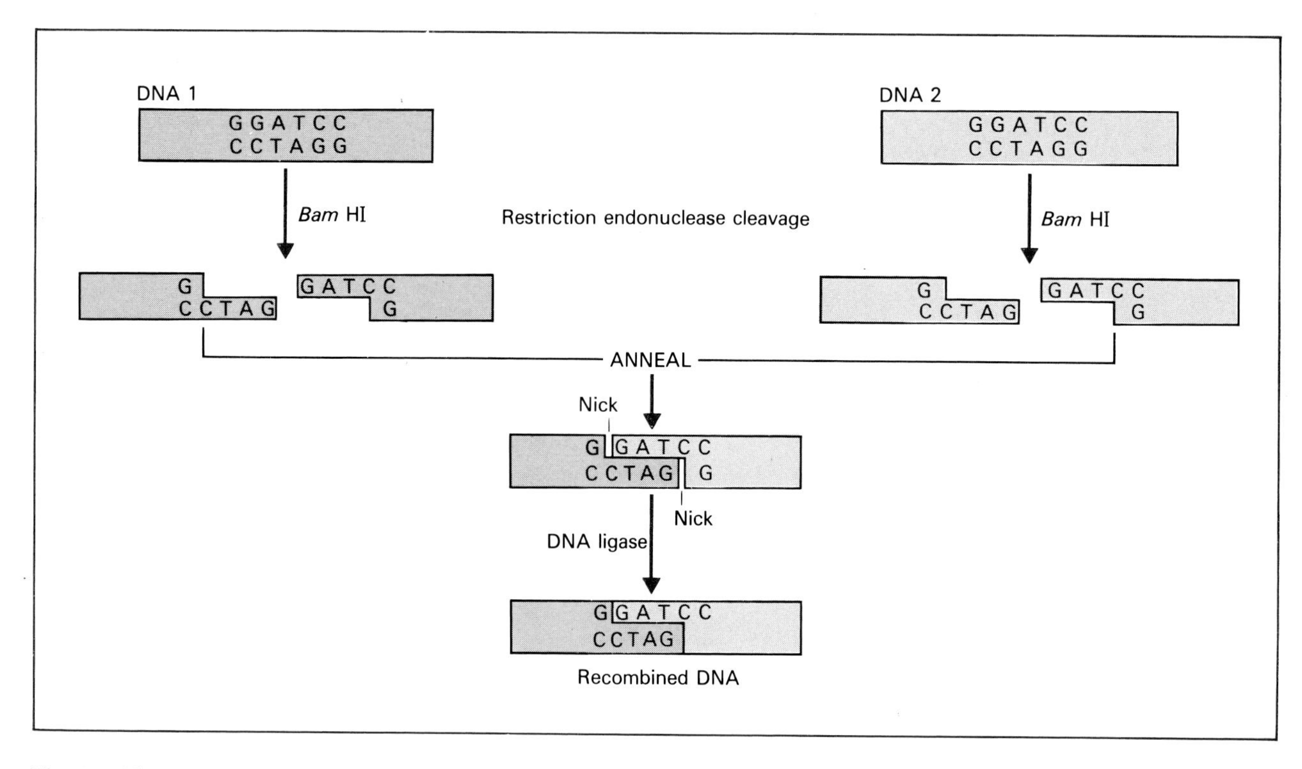

Fig. 2.3 The construction of a chimaeric (or recombinant) DNA molecule by joining together two DNA fragments produced by cleavage of different parental DNA molecules with the same restriction endonuclease.

PLASMIDS AS CLONING VECTORS

Hundreds of different cloning vectors, of which a representative selection is shown in Fig. 2.4, have been described and many of them have been constructed for special purposes. However all of them have three features in common.

1 The DNA of the cloning vector ideally has only a single target site for any particular restriction endonuclease. If it has more than one target site, bits of the plasmid can be lost or rearranged during the cloning process. The more restriction endonucleases for which the plasmid has unique sites then the better it will be as a vector.

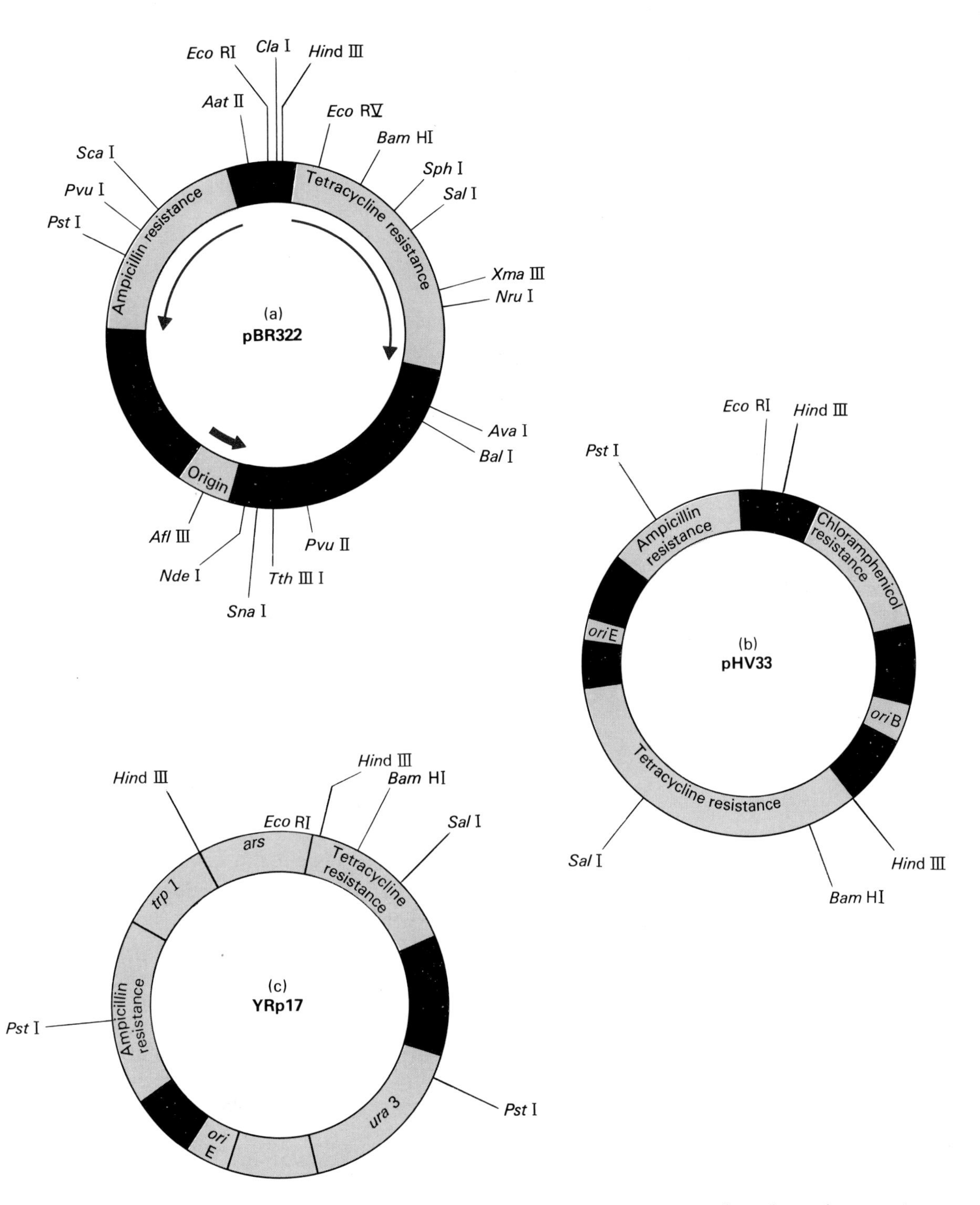

Fig. 2.4 Three plasmid vectors used for gene cloning (not drawn to scale). Restriction endonuclease cleavage sites are indicated around the perimeter of each vector. (a) The structure of pBR322, a vector used with *E. coli*. The thin arrows inside the circle show the direction of transcription of the two antibiotic resistance genes and the thick arrow the direction of DNA replication. (b) The structure of pHV33, a plasmid vector which can replicate in both *E. coli* and *B. subtilis*. *ori*E refers to the origin of replication necessary for plasmid maintenance in *E. coli* and *ori*B that for maintenance in *B. subtilis*. (c) A cloning vector which can replicate in both *E. coli* and yeast. The *ars* element is essential for replication in yeast. *trp* 1 and *ura* 3 are two yeast genes involved in the biosynthesis of tryptophan and uracil respectively. *ars* is a DNA sequence conferring autonomous replication of the vector in yeast cells.

2 All useful cloning vectors have one or more readily selectable genetic marker such as antibiotic resistance. The efficiency of the transformation process is very low and such markers are essential so that cells which have been transformed can be selected, e.g. on the basis of acquisition of ampicillin resistance.
3 As indicated above, one property of vectors is their possession of an origin of replication which ensures that they are propagated in the desired host cell. In some instances a plasmid is used as a cloning vector in two unrelated host cells, e.g. *Escherichia coli* and *Bacillus subtilis.* Such bifunctional vectors have more than one origin of replication (Fig. 2.4). Some cloning vectors used in yeast carry a centromere to facilitate segregation at cell division.

EXPRESSION OF CLONED GENES

Theoretically it is possible to clone in one organism any desired gene from another organism by the technique known as *shotgunning.* To do this the entire genome of the first organism is digested with a restriction endonuclease to produce a random mixture of fragments. These fragments are inserted into a plasmid vector and the recombinant plasmids transformed into the desired host cell. Since each recombinant plasmid will contain a different fragment of foreign DNA it is a major task (see p. 19) to select those transformed cells that carry the cloned gene of interest, in this case the human insulin gene. Even when the right cells can be identified, it is highly probable that they would not synthesize human insulin. That is, the information contained in the cloned gene is not used to create a functional protein or, in common parlance, the gene is not *expressed.*

There are a number of steps in gene expression:
1 transcription of DNA to mRNA;
2 translation of the mRNA into a polypeptide sequence;
3 in some instances, post-translational modification of the protein.
In many eukaryotes there is an additional step before translation, the removal of non-coding sequences from mRNA by a process known as splicing.

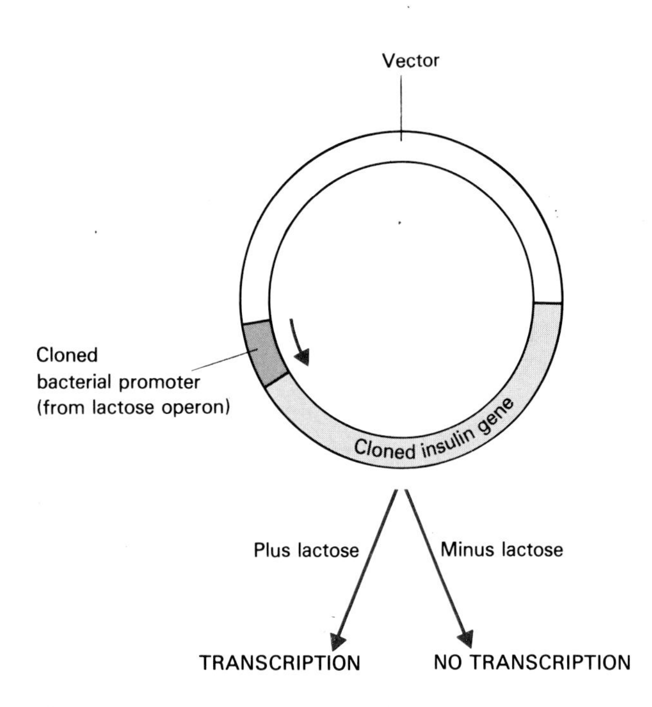

Fig. 2.5 Insertion of a cloned insulin gene into a vector carrying a bacterial promoter. The arrow within the circle indicates the direction of transcription. If the promoter is derived from the lactose operon, the transcription will be initiated only in the presence of lactose or a lactose analogue.

Transcription Transcription of DNA into mRNA is mediated by the enzyme RNA polymerase. The process starts with RNA polymerase binding to recognition sites on the DNA which are called *promoters.* After binding, the RNA polymerase molecule travels along the DNA molecule until a termination signal is encountered. It follows that a gene which does not lie between a promoter and a termination signal will not be transcribed. Genes isolated in certain ways, such as by cDNA cloning or artifical synthesis (see below) do not have their own promoter and they must be inserted into a vector close to a promoter site. An example is shown in Fig. 2.5. Even if a cloned gene carries its own promoter, this promoter may not function in the new host cell. In such circumstances the original promoter has to be replaced.

Translation Translation of mRNA into protein is a complex process which involves interaction of the messenger with ribosomes. For translation to take place the mRNA must carry a *ribosome binding site* (rbs) in front of the gene to be translated. After binding, the ribosome moves along the mRNA and initiates protein synthesis at the first AUG codon it encounters and continues until it encounters a stop codon (UAA, UAG or UGA). If the cloned gene lacks a ribosome binding site, it is necessary to use a vector in which the gene can be inserted downstream from both a promoter and an rbs (Fig. 2.6).

Splicing mRNA Genes from bacteria and viruses have a very simple structure in that all the genetic information in the mRNA between the initiation and stop codons is translated into protein. Many genes of eukaryotic organisms, including the human insulin gene, have a more complex structure. They are a mixture of coding regions or *exons*, which contribute to the final protein sequence, and non-coding regions or *introns*, which are not translated into protein. In eukaryotes, genes containing introns are transcribed into mRNA in the usual manner but then the corresponding intron sequences are spliced out (Fig. 2.7). As bacteria cannot splice out introns they cannot be used directly to express many genes from mammals or other eukaryotes.

One solution to the problem of introns is to clone the gene of interest in yeast (*Saccharomyces cerevisiae*), which can mediate splicing. Unfortunately cloning directly into yeast is much less efficient than cloning into a bacterium such as *E. coli* and

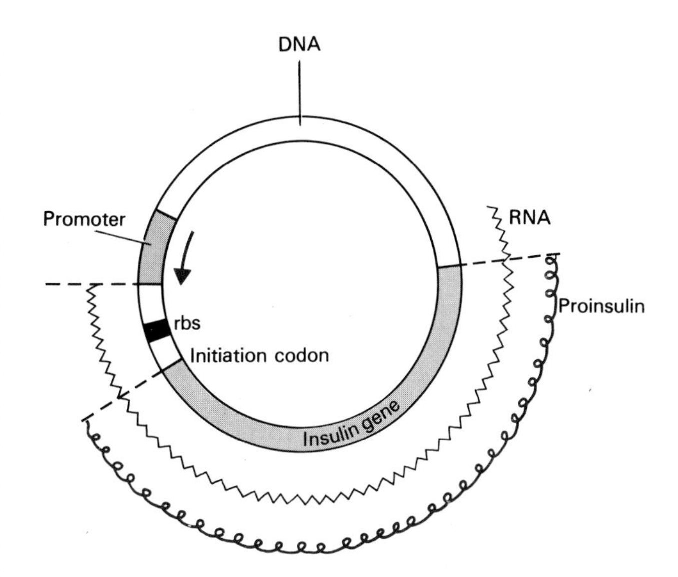

Fig. 2.6 The use of a vector carrying a promoter, an adjacent ribosome binding site (rbs) and an initiation codon to obtain synthesis of proinsulin from a synthetic gene. The arrow indicates the direction of transcription.

this technique is not favoured. A better approach is to start by isolating mRNA from the original organism. In the case of insulin the mRNA would be obtained from human pancreatic cells, as these cells

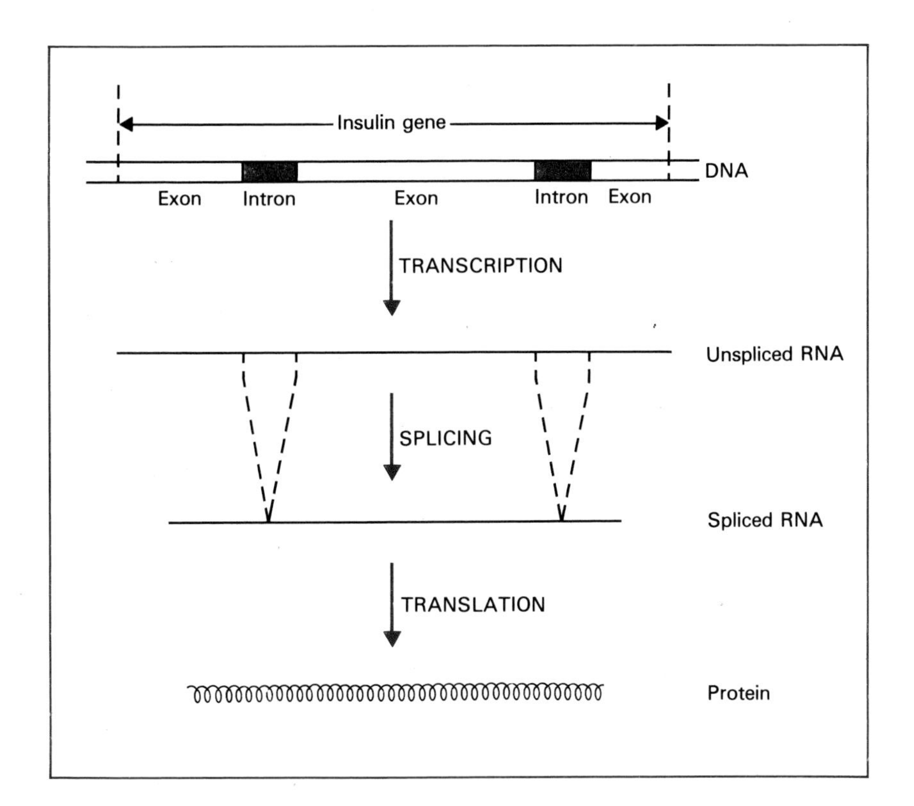

Fig. 2.7 Splicing of a messenger RNA molecule transcribed from a hypothetical insulin gene containing two introns.

are rich in insulin-specific mRNA from which introns have already been spliced out. Using the enzyme *reverse transcriptase* it is possible to convert this mRNA into a DNA copy (Fig. 2.8). This copy DNA (*cDNA*), which carries the uninterrupted genetic information for insulin production, can then be cloned. It should be noted that such cDNA has *blunt ends* instead of sticky ends but it still can be inserted into plasmid vectors using blunt-end ligation (Fig. 2.9).

An alternative way of avoiding the problem of introns is to synthesize an artificial gene in the test tube starting with deoxyribonucleotides (see Box, p. 18). In practice various oligonucleotides are synthesized and ligated together before insertion into an appropriate vector. This approach has been used to clone genes encoding proteins up to 500 amino acids long but it demands that the entire amino acid sequence be known in advance.

Post-translational modifications A number of proteins undergo post-translational modifications and insulin is one of these. Proteins that are destined to be transported out of the cell are synthesized with an extra 15–30 amino acids at the amino-terminus (N-terminus). These extra amino acids are referred to as a *signal sequence* and a common feature of these sequences is that they have a central core of hydrophobic amino acids flanked by polar or hydrophilic residues. During passage through the membrane the signal sequence is cleaved off (Fig. 2.10). If our insulin gene were cloned by the cDNA method, the signal sequence would be present and, in *E. coli* at least, the insulin would be transported through the cytoplasmic membrane (*exported*). Using the synthetic gene approach a signal sequence would be present on the protein only if the nucleotide sequence corresponding to the protein's signal sequence had been incorporated at the time of gene

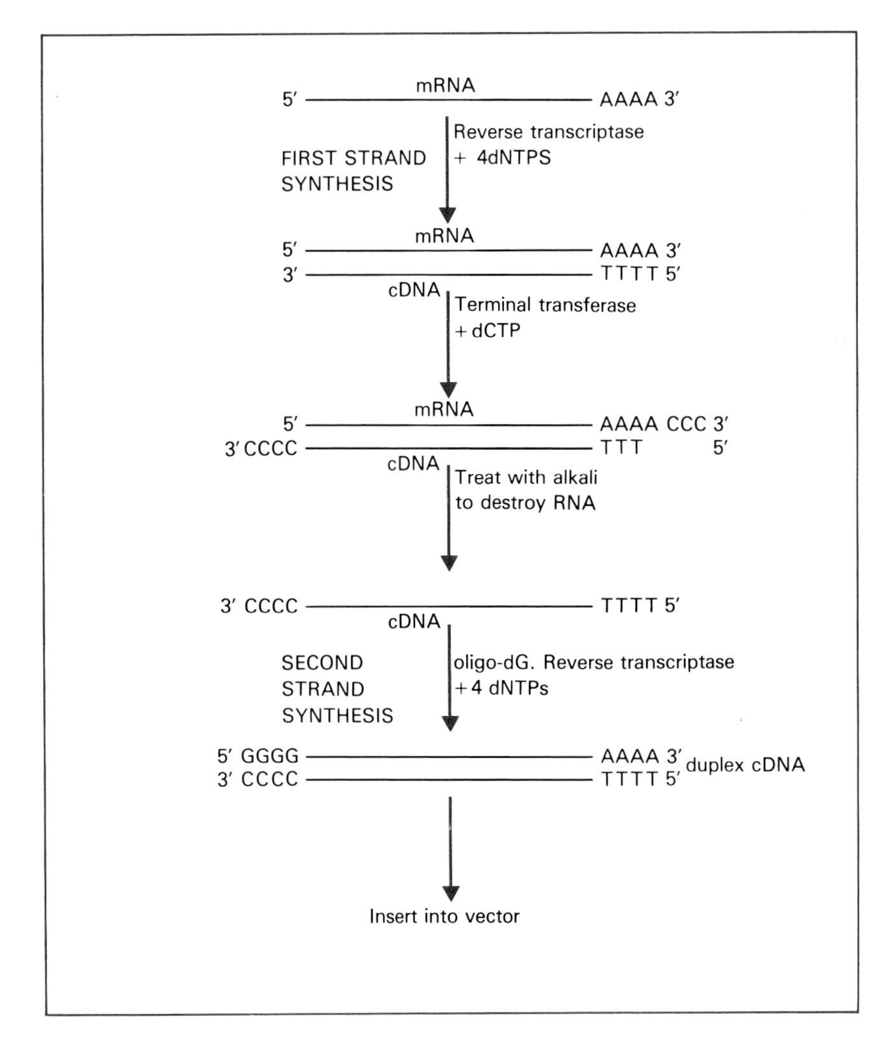

Fig. 2.8 Simplified method of using reverse transcriptase to convert mRNA into double-stranded DNA suitable for cloning.

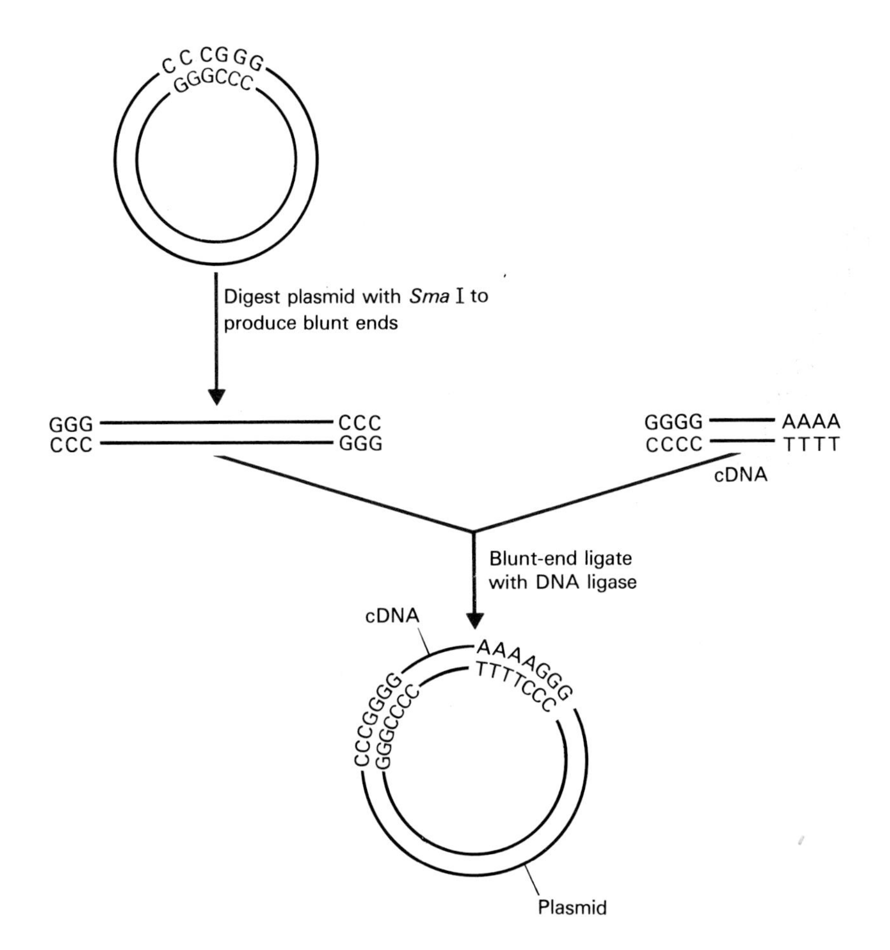

Fig. 2.9 The use of blunt-end ligation to clone a cDNA fragment. The DNA sequence shown on the plasmid is the recognition site for restriction endonuclease *Sma* I (see Fig. 2.2). The cDNA shown is the one generated in Fig. 2.8.

construction. Sometimes it is desirable for the bacterium to export the protein, in which case a signal sequence is incorporated; in other cases it may be desirable that proteins are retained within the cell.

In *E. coli* cells the presence of a signal sequence usually does not result in export of a protein into the growth medium. Instead it is directed to the periplasmic space, that is the space between the cytoplasmic membrane and the complex membrane-like cell wall. Unfortunately, many recombinant proteins are rapidly and extensively degraded in the periplasmic space due to the presence there of numerous proteolytic enzymes (*proteases*). In Gram-positive species such as *Bacillus* and in eukaryotic microorganisms signal sequences direct the export of proteins into the growth medium.

The export process is even more complex in that protein synthesis and export occur simultaneously on the inner surface of the cell membrane. If the rate of synthesis of an exported protein is very high, it is possible to envisage a situation in which the membrane becomes 'jammed' with protein in the various stages of synthesis and export. This problem can be minimized by using as a host cell for the recombinant plasmid one which has a very high surface area to volume ratio. Suitable organisms are actinomycetes and filamentous fungi.

A small number of proteins, and again insulin is a good example, are synthesized as *proproteins*. This means that there is an additional amino acid sequence which dictates the final three-dimensional structure but which is deleted before the protein becomes fully functional (Fig. 2.10). In the case of proinsulin, proteolytic attack cleaves out a stretch of 35 amino acids in the middle of the molecule to generate insulin. The peptide that is removed is known as the C fragment. The other chains, A and B, remain cross-linked and thus locked in a stable tertiary structure by the disulphide bridges formed when the molecule originally folded as proinsulin. Bacteria have no mechanism for specifically cutting out the folding sequences from proproteins. If native

The chemical synthesis of oligonucleotides and genes

The ability to synthesize genes chemically is fundamental to the success of modern biotechnology. Not only does it enable complete genes to be synthesized to facilitate cloning, as described in this chapter, but it enables minor or major changes to be made in the gene product. This latter aspect, protein engineering, is discussed in Chapter 4. In addition, short oligonucleotides, which are easy to synthesize, can be used diagnostically in a variety of ways as described in Chapter 3.

The basic method of gene synthesis is the repetitive formation of an ester linkage between an activated phosphoric acid function of one nucleotide and the hydroxyl group of another nucleoside or nucleotide thus forming the characteristic phosphodiester bridge. The major problem is that deoxyribonucleotides are very reactive molecules, having a primary and secondary hydroxyl group, a primary amino group and a phosphate group. Consequently, blocking and deblocking procedures are required and the chemistry involved must not result in scission or alteration of the phosphodiester backbone, the furanose rings, the sugar–purine/pyrimidine bond or the bases themselves — a tall order indeed!

Currently, the most popular method for the synthesis of short nucleotides (called *oligonucleotides*) makes use of the extreme reactivity of phosphite reagents. The basic principles of the method are shown in Diagram 1. Although the method will work in solution for the condensation of 3 or 4 nucleotides, construction of large oligonucleotides requires the 3′ end of the desired oligonucleotide to be coupled to an insoluble support. Fortunately, immobilization simplifies manipulative procedures and cuts out time-consuming purification steps following each cycle of condensation. Appropriately blocked mononucleotides are added sequentially and reagents, starting materials and by-products removed by filtration. At the conclusion of the synthesis the deoxyoligonucleotide is chemically freed of blocking groups, hydrolysed from the support and purified by electrophoresis or high performance liquid chromatography (HPLC). By immobilizing the polymer support carrying the initiating deoxynucleotide in a column, the filtration steps can be replaced by a simple washing procedure and this lends itself to a fully automatic synthesis. Automatic gene synthesizers consist simply of reagent reservoirs whose contents are added to or removed from the immobilized protected oligonucleotide via valves controlled by a microprocessor.

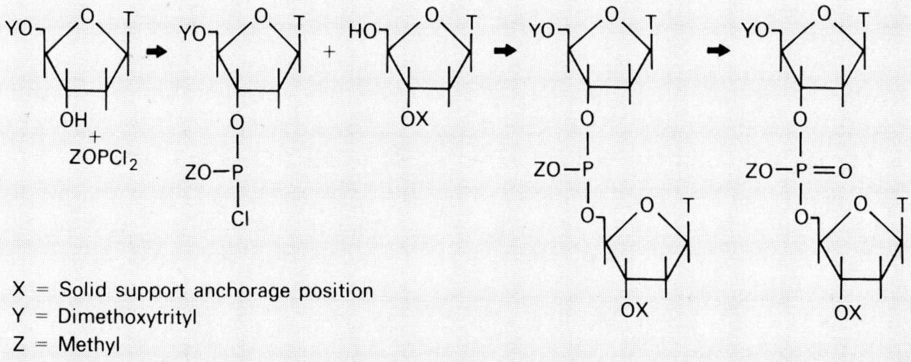

X = Solid support anchorage position
Y = Dimethoxytrityl
Z = Methyl

Diagram 1

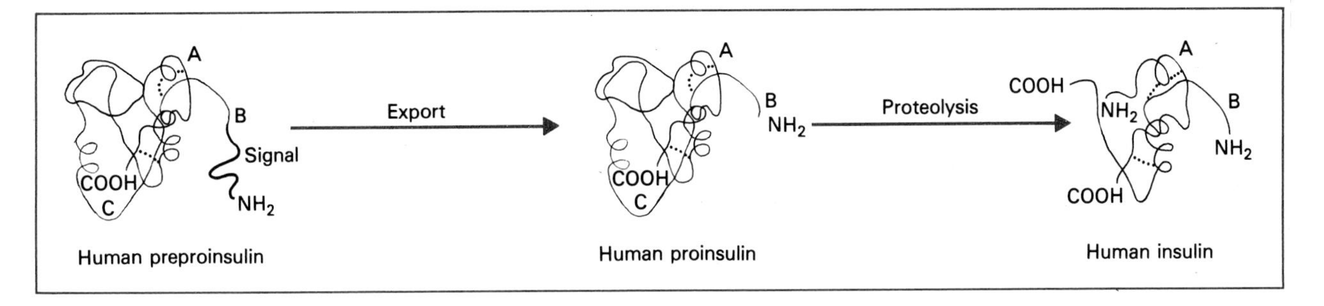

Fig. 2.10 The conversion of preproinsulin to insulin by sequential removal of the signal peptide and the C fragment.

insulin is to be made using recombinant bacteria, it is necessary to synthesize the A and B chains separately, purify them and join them together *in vitro*.

Another post-translational modification occurring naturally in the construction of many proteins is the addition of oligosaccharides to certain amino acid residues. This process of *glycosylation* occurs, for example, in the synthesis of the antiviral interferon-β and -γ. Bacteria cannot glycosylate the products of cloned mammalian genes. Non-glycosylated proteins usually retain their pharmacological or biological activities but their stability *in vitro* and *in vivo* and their distribution in the animal body may be different. Yeast cells can glycosylate proteins but the pattern of glycosylation differs from that mediated by animal cells. Where the correct glycosylation is necessary the only option is to clone the corresponding gene in a mammalian cell (see Chapter 9).

SELECTION OF RECOMBINANTS

The task of isolating a desired recombinant from a population of transformed bacteria depends very much on the cloning strategy that has been adopted; for instance, if a synthetic gene has been cloned, no selection is necessary because every transformed cell will contain the correct sequence. When a cDNA derived from a purified or abundant mRNA is to be cloned, the task is relatively simple: only a small number of clones needs to be screened. Isolating a particular single-copy gene sequence from a complete mammalian genomic library requires techniques in which hundreds of thousands of recombinants can be screened. The advantages and disadvantages of the different cloning strategies are shown in Table 2.1.

A number of different methods have been devised to facilitate screening of recombinants. These include genetic methods, immunochemical methods and methods based on nucleic acid hybridization. The simplest example of a genetic method is the complementation of nutritional defects; for example, suppose a bacterial strain is available which has a mutation in a gene encoding an enzyme involved in the biosynthesis of the amino acid histidine. Such a mutant strain will only grow in medium supplemented with histidine. By cloning DNA from a normal strain, i.e. one that can synthesize its own histidine, in the mutant strain and selecting those transformants which grow in the absence of histidine it is possible to isolate the gene of interest. Even yeast and mammalian genes can be selected in this way.

In a procedure analogous to that described above, a clone carrying the mouse dihydrofolate reductase (DHFR) gene was selected from a population of recombinant plasmids containing cDNA derived from an unfractionated mouse cell mRNA preparation. In this instance the basis of selection was that mouse DHFR is more resistant to the inhibitor trimethoprim than *E. coli* DHFR. When trimethoprim was added to the medium on which transformed *E. coli* were grown, only those cells carrying the mouse DHFR gene survived.

Immunochemical methods are widely used to select particular recombinant clones but demand the availability of a specific antibody against the desired protein product. When such an antibody is available the method works well. Transformed cells are grown on agar in a conventional Petri dish and a duplicate set of colonies prepared in a second Petri dish. The cells in the duplicate set of colonies are lysed by exposure to chloroform vapour and the

Table 2.1 Comparison of the different cloning strategies.

Cloning method	Advantages	Disadvantages
Shotgun cloning	1 Easy to do	1 Many different clones isolated so need good selection method to isolate desired clone 2 Genes with introns will not be correctly expressed in *E. coli*, the usual shotgun cloning host 3 Expression dependent on recognition of foreign promoters by *E. coli* 4 Codon choice may not be optimal for *E. coli*
cDNA cloning	1 If desired mRNA is in abundance in the mRNA preparation used, desired clone should be easily detected 2 Introns not a problem as mRNA will contain spliced molecules	1 Desired mRNA not always abundant so selection may be necessary 2 Technically more difficult than shotgun cloning 3 Cloned gene needs to be placed downstream from a promoter, since normal promoter eliminated by starting with mRNA 4 Codon choice may not be optimal for *E. coli*
Gene synthesis	1 No selection needed after cloning 2 Sequence of promoters, ribosome binding sites, etc., can be optimized 3 Codon choice can be optimized for preferred host cell	1 Need to know protein sequence before a synthetic gene can be designed

released proteins blotted onto an adsorbent matrix. In effect, this produces a map of the original colonies in the form of their protein products. The matrix is then exposed to antibody which has been radioactively labelled *in vitro*. Positively reacting lysates are detected by washing surplus radiolabelled material off the matrix and making an autoradiographic image (Fig. 2.11). Many different variations of this technique have been adopted.

Detection methods based on nucleic acid hybridization are also widely used but these techniques have other applications as well and are described in more detail in Chapter 3.

MAXIMIZING GENE EXPRESSION

If the objective of cloning a mammalian gene in a microorganism is to facilitate commercial production of the corresponding gene product, it follows that it is essential to maximize gene expression. Important factors are:

1 the number of copies of the plasmid vector per unit cell (*copy number*);
2 the strength of the promoter;
3 the sequence of the rbs and flanking DNA;
4 codon choice in the cloned gene;
5 genetic stability of the recombinant;
6 proteolysis.

The limiting factor in expression is the initiation of protein synthesis. Increasing the number of plasmids per cell increases the number of mRNA molecules transcribed from the cloned gene and this results in increased protein synthesis. Similarly, the stronger the promoter, the more mRNA molecules

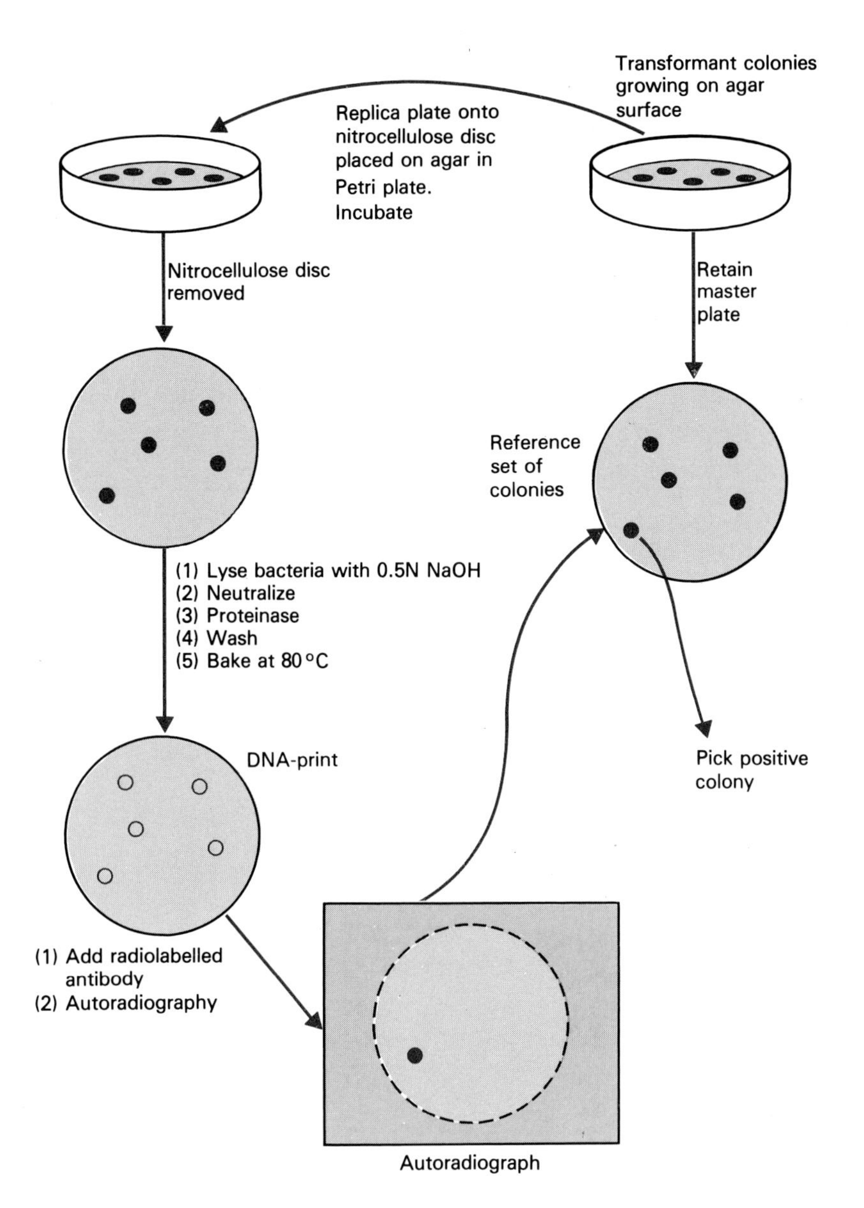

Fig. 2.11 The basic method of immunological screening of recombinants.

are synthesized. The nucleotide base sequence of the rbs and the length and sequence of the DNA between the rbs and the initiating AUG codon are so important that a single base change, addition or deletion can affect the level of translation up to 1000-fold.

Another important factor is related to the redundancy of the genetic code. There are several trinucleotide codons for most amino acids and different organisms favour different codons in their genes. If genes inserted into cells of another species utilize codons rare in the host cell, the host's biosynthetic machinery may be starved of charged tRNAs. This could result in premature protein chain termination or a high error frequency in the amino acid sequence of the protein.

In any culture containing cells engaged in excess synthesis, be it of a protein or a low molecular weight molecule (e.g. an antibiotic), there is a strong selective pressure in favour of non-producing cells. Cells which are not encumbered by extra production can invest their energy in cell division and hence rapidly outnumber producers. Non-producing cells frequently arise in transformed cultures as a result of spontaneous changes in gene sequence, e.g. deletions and gene rearrangements. To minimize selection against cells which can achieve high product levels it is wise to minimize recombinant gene expression until the final production vessel is reached. This can be achieved by using controllable promoters and plasmid vectors with controllable copy number. An example of a controllable promoter is that derived from the *E. coli* lactose operon: no transcription occurs unless lactose is added (Fig. 2.5). *Runaway plasmids* are examples of vectors with controllable copy number. At low temperatures, e.g. below 30 °C, the copy number may be as low as 10 plasmids per cell but when the temperature is raised to 37 °C the copy number increases to several hundred, or even several thousand.

The enzymatic breakdown of protein (*proteolysis*) does not affect transcription and translation but by degrading the desired product it influences the *apparent* rate of gene expression. Although proteolysis can be reduced, it is difficult to eliminate completely. One approach which is used widely is to 'protect' the desired protein by fusing it to a normal cellular protein from which it must subsequently be released. This approach is discussed in more detail in Chapter 4.

Further reading

GENERAL

Old R.W. & Primrose S.B. (1985) *Principles of Gene Manipulation*, 3rd edn. Blackwell Scientific Publications, Oxford.

SPECIFIC

Holland I.B., Mackman N. & Nicaud J-M. (1986) Secretion of proteins from bacteria. *Bio/technology* **4**, 427–31.

Hsuing H.M., Mayne N.G. & Becker G.W. (1986) High level expression, efficient secretion and folding of human growth hormone in *Escherichia coli*. *Bio/technology* **4**, 991–5.

Kaplan B.E. (1985) The automated synthesis of oligodeoxyribonucleotides. *Trends in Biotechnology* **3**, 253–6.

Reznikoff W. & Gold L.R. (eds) (1986) *Maximizing Gene Expression*. Butterworths, London.

3/Gene Probes as Diagnostic Reagents and Molecular Fingerprints

INTRODUCTION

The production of reagents for the diagnosis of biological disorders and infectious disease is a major industry and on a global scale revenue from sales is estimated at many billions of dollars. Traditionally diagnostic reagents have taken two forms:

1 biochemical reagents for assaying specific enzymes;

2 antibodies for detecting specific proteins by such techniques as immunofluorescence, radioimmunoassay (RIA) and enzyme-linked immunosorbent assay (ELISA).

Although the former are used principally in clinical biochemistry, the latter have applications in almost all industries, e.g. health care, food and agriculture. Recently a new class of diagnostic reagent has been developed which relies on the use of DNA or RNA probes to detect particular gene sequences. As will be discussed later, applications of the technique include the identification of hereditary defects in

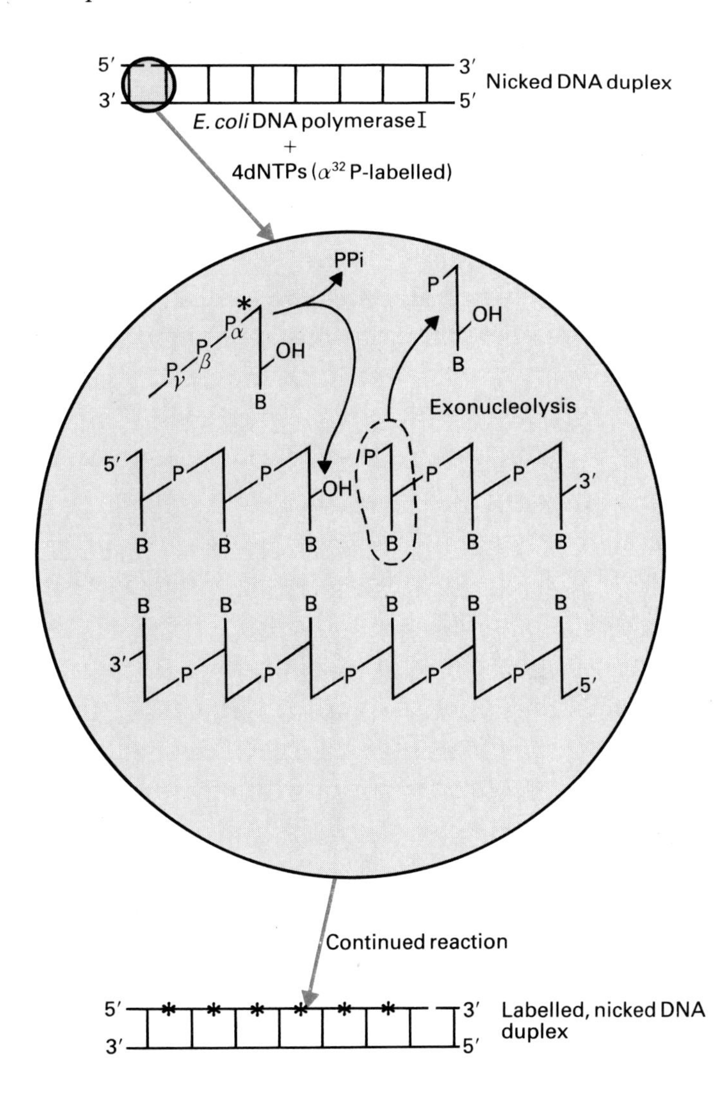

Fig. 3.1 Synthesis of labelled DNA by nick translation.

embryos and the identification of particular plant diseases. Although based on an old technique, nucleic acid hybridization, development of these reagents awaited the availability of milligram quantities of specific genes. The ready provision of such large amounts of gene sequences is easy if the sequences have been cloned in a vector, for a 1 l bacterial culture can yield up to half a milligram of plasmid DNA.

For hybridization of a nucleic acid probe to a test gene to be detected or quantified some method of labelling the probe is required. In the past, probes with a radioactive label were prepared from bacteria grown in the presence of radioisotopes. More recently, methods of labelling probes *in vitro* have been devised.

LABELLING NUCLEIC ACIDS *IN VITRO*

One method of labelling DNA gene probes *in vitro* is by means of a technique known as *nick translation* (Fig. 3.1). In any preparation of DNA, particularly one that is more than a few days old, random single-strand breaks ('nicks') occur. Starting at the free 3′-hydroxyl group exposed in these nicks, the enzyme DNA polymerase I (see adjacent Box) will incorporate free nucleotides. Concomitantly the enzyme digests away single-stranded DNA from the exposed 5′ end. If one or more of the deoxynucleoside triphosphates is labelled in the α position with phosphorus-32, the reaction progressively incorporates the radioactive label into a duplex that is unchanged except for movement ('translation') of the nick along the molecule. DNA probes of high specific activity (10^8 disintegrations per minute (dpm)/μg DNA) can be obtained readily by this method.

An alternative method of labelling DNA is to use the *random primer method*. The basic principle of the method is shown in Fig. 3.2. A mixture of random-sequence hexanucleotides is added to DNA. Some of those hexanucleotides will hybridize to the DNA and act as initiation sites for DNA polymerase. When the four deoxynucleotide triphosphates are added, only one of which need be radioactively labelled, probes 200–400 nucleotides in length are produced. These probes can be released by denaturation and used for hybridization without purification. By this method, probes can be produced

DNA polymerase

A number of DNA polymerases from eukaryotic and prokaryotic sources have been well characterized. Most of those from eukaryotes possess only a DNA synthesizing activity, whereas the corresponding bacterial enzymes also possess exonuclease activity. There are three DNA polymerases in *E. coli*, of which DNA polymerase I is the most versatile and useful. DNA polymerase I is a single polypeptide of molecular weight 109 000 and possesses three activities: a DNA synthesizing activity, a 3′ → 5′ exonuclease activity and a 5′ → 3′ exonuclease activity. The enzyme can be cleaved into two fragments by protease treatment. The larger fragment, known as the *Klenow fragment*, has a molecular weight of 76 000 and contains the polymerase and 3′ → 5′ exonuclease activities. The small fragment (molecular weight 36 000) has the 5′ → 3′ exonuclease activity.

DNA polymerase I has the unique ability to initiate replication *in vitro* at a nick in DNA

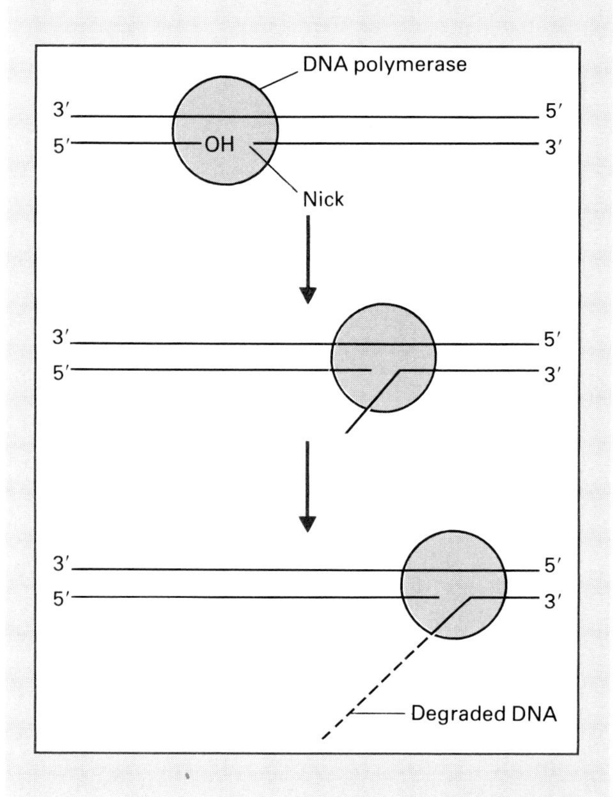

Diagram 1 Nick translation of DNA.

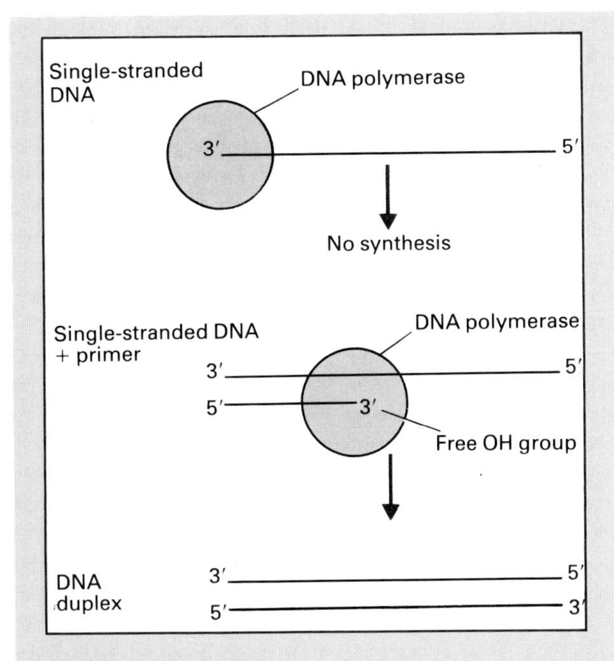

Diagram 2 The need for a primer for DNA synthesis.

and to extend the 3′-OH end. As the new segment of DNA is synthesized, it displaces the existing homologous strand in the duplex. The displaced strand, in turn, is digested by the 5′ → 3′ exonuclease activity. This process of *nick translation* (Diagram 1) is used to radiolabel DNA *in vitro* (see main text).

All of the known DNA polymerases can extend a DNA chain only from a free 3′-OH end. When presented with a single-stranded template none can initiate synthesis unless a start primer is added to present the requisite 3′-OH group (Diagram 2). These primers may be either RNA or DNA.

whose specific activity is ten times higher than those prepared by nick translation.

Single-stranded RNA can also be used as a gene probe. Indeed, it is probably the most sensitive probe for any diagnostic procedure based on hybridization because the rate and stability of hybridization is far greater in an RNA–DNA hybrid than in a DNA–DNA hybrid. The best system for labelling RNA is the SP6 system (Fig. 3.3) since it readily yields specific activities in excess of 5×10^9 dpm/μg RNA. The gene from which the probe is to be prepared is cloned in a plasmid vector such that it is under the control of the promoter of the phage SP6 RNA polymerase gene. After purification the plasmid is linearized with a suitable restriction enzyme and then incubated with SP6 RNA polymerase and the four ribonucleoside triphosphates. If one of the triphosphates is radioactively labelled, the incubation mix will contain labelled RNA, which can be added directly to the hybridization mixture.

Example: Detection of viroids in plants Viral infection of crop plants can have serious agronomic consequences, not the least of which is decreased vigour and reduced crop yield. Preventive measures include the cultivation of crops from virus-free stocks and the prevention of these pathogens from entering and spreading through crops. In these procedures suitable diagnostic tests for the rapid and reliable detection of viruses are of paramount importance. Enzyme-linked immunosorbent assays (ELISA) have been developed for the detection of a number of important plant viruses and many of these are in widespread use. Viroids cause a number of important plant diseases (see Box below) but because they lack the antigenic protein coat charac-

Viroids

Viroids are responsible for diseases of a number of important crop and ornamental plants including potatoes (potato spindle tuber viroid), chrysanthemums (stunt and chlorotic mottle viroids), hops (stunt viroid), citrus (exocortis viroid) and avocado (sunblotch viroid). Unlike viruses they are never found encapsidated within protein but always exist as naked nucleic acid. All are similar, consisting of a single-stranded circular RNA molecule. They are the smallest known infectious agents (mol. wt 1.2×10^5) and they do not appear to code for any protein. The genome of potato spindle tuber viroid has been completely sequenced and part of it bears a striking similarity to sequences found in regions of RNA associated with intron removal (see Fig. 2.7). This has led to the suggestion that viroids may have originated as 'escaped introns'. If this is so, their pathogenicity may be due to interference with normal RNA splicing.

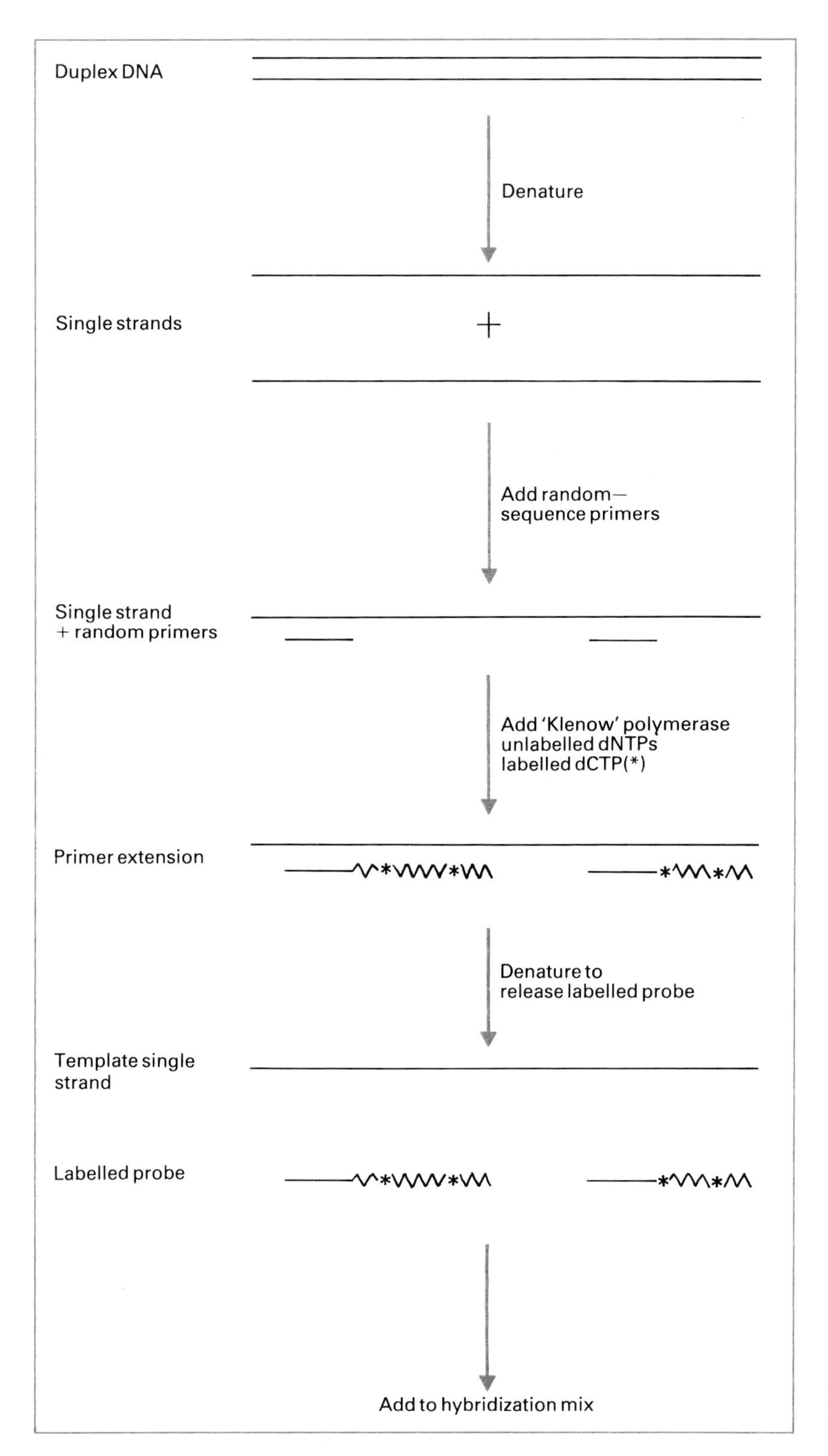

Fig. 3.2 The random primer method for preparing labelled DNA.

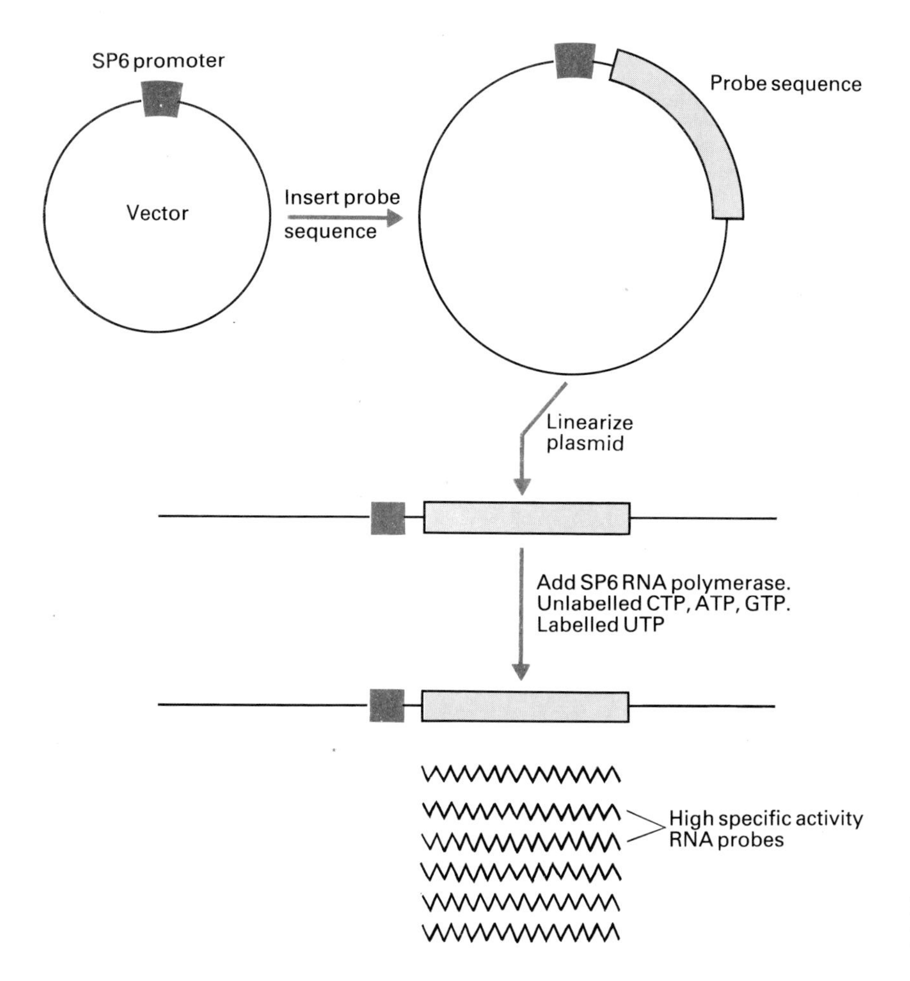

Fig. 3.3 The SP6 system for preparing labelled RNA.

teristic of viruses they cannot be detected by immunological procedures. However, they can be detected by probing plant extracts with a radioactively labelled plasmid carrying a copy of the viroid genome. The method is shown in Fig. 3.4.

Test plants are homogenized in a buffer and the resulting extract clarified by centrifugation. The clarified extract is spotted onto a nitrocellulose membrane and dried in a vacuum which causes the viroid RNA to bind irreversibly. Binding occurs by means of the sugar-phosphate backbone of the RNA and leaves the bases available for hydrogen bonding to complementary bases. An immediate problem arises in that added probe DNA will also bind to the filters. To avoid this problem any nucleic acid binding sites remaining on the nitrocellulose after the test RNA has bound are blocked with bovine serum albumin. When the filters have been prepared the radiolabelled probe is added and the mixture incubated under conditions appropriate for hybridization. The filter is washed to remove excess probe which has not hybridized and any positive samples are identified by autoradiography.

Probes of the kind described above can also be used to facilitate plant breeding; for example, one goal of potato breeding programmes is the development of cultivars which are resistant to virus diseases transmitted by aphids. Identification of suitable plants is effected by exposing them to virus-bearing aphids and then testing the cell sap with DNA probes for potato viruses X and Y and potato leaf-roll virus. A different application is in wheat breeding. Many European wheats have a chromosome IB in which the short arm is derived from rye. Some breeders are interested in selecting plants in which this chromosome is entirely from wheat. By using a rye DNA sequence as a probe, plants with IB^{s} from rye can be detected simply by squashing a root tip onto a nitrocellulose filter and then applying a suitable probe.

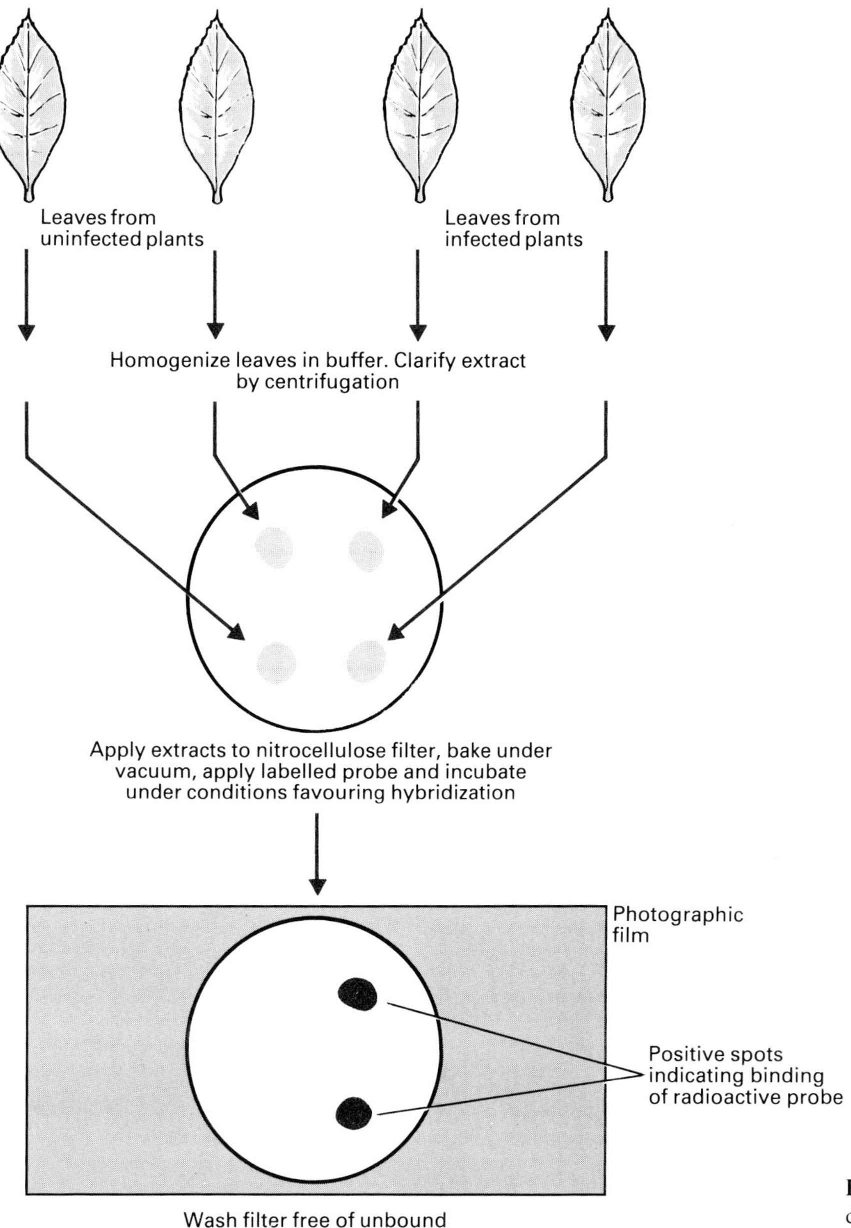

Fig. 3.4 Use of a gene probe to detect plants infected with a viroid. See text for details.

USE OF GENE PROBES VERSUS IMMUNOLOGICAL DETECTION

The use of gene probes is not restricted to the detection of viroids; with a slight modification they can also be used to detect bacteria. In this case the test sample is applied to a nitrocellulose filter as before. If the filter is dropped into sodium hydroxide solution, the bacteria will lyse, the released DNA will denature and the separated strands will bind loosely to the nitrocellulose. After washing to remove the alkali, the filter is dried in a vacuum oven as before. Probing is then carried out as described in the previous section.

Although a large number of different micro-organisms can now be detected using gene probes, the method has two disadvantages when compared with immunological methods. The first of these is the use of radioactive labels, although non-radioactive methods of labelling nucleic acids are being developed and these are described in the next section. The second disadvantage is the overall test time, which is usually 24–72 h. With the use of non-radioactive labelling methods this may be reduced to 6–12 h but even this is slower than a serological method.

On the positive side, probe technology has two advantages over immunological detection methods. First, the method can be applied without modification to samples of blood, faeces, pus or body exudates which are too dirty to permit antibody–antigen interaction. The alkali denaturation step effectively cleans the sample. Second, probe technology can detect pathogenicity determinants which would not be revealed immunologically; for example, many of the cases of diarrhoea and vomiting which occur amongst persons travelling abroad, the so-called 'travellers tummy', are caused by strains of *Escherichia coli* which secrete an enterotoxin. Whereas the genes for enterotoxin are plasmid-borne and are detected easily with appropriate probes, there are no differences between toxigenic and non-toxigenic *E. coli* which can be detected with antisera.

NON-RADIOACTIVE LABELS FOR PROBES

Although probes labelled with radioactive phosphorus are convenient for the research worker, they are not particularly suitable for a diagnostic laboratory where quality control and long shelf-life are important for standardization of tests. By using radioactively labelled thionucleotides the shelf-life of the probes can be extended. An alternative solution is to use biotin-labelled nucleoside triphosphates during nick translation. Binding of the biotin-labelled DNA is detected by means of an avidin/alkaline phosphatase system (Fig. 3.5). Various enhancement techniques are being developed to increase the sensitivity of this system.

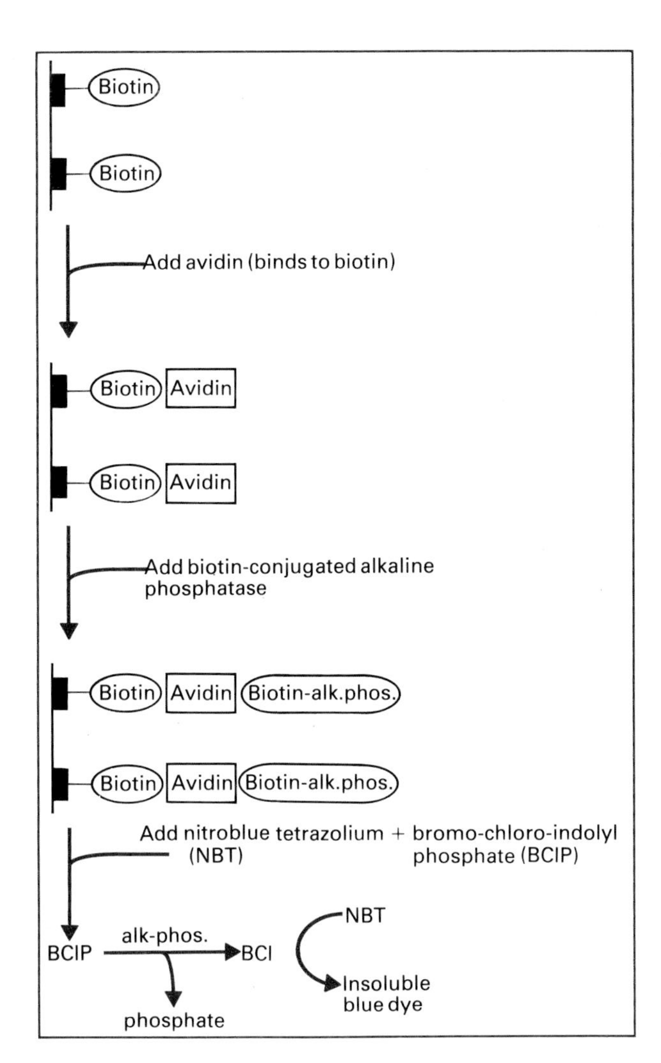

Fig. 3.5 The detection of biotin-labelled DNA using avidin and alkaline phosphatase. The insoluble blue dye produced by the reduction of nitroblue tetrazolium 'stains' the nitrocellulose in the vicinity of biotin-labelled DNA.

SOUTHERN BLOTTING

The diagnostic potential of DNA–DNA hybridization can be increased greatly by means of *Southern blotting*. This technique is named after Ed Southern who developed it in 1975. The basic principle of the method is shown in Fig. 3.6. A sample of DNA is digested with a restriction endonuclease and the fragments which are produced are separated on a size basis by electrophoresis in an agarose gel. The DNA fragments are then denatured into single strands by soaking the gel in alkali. Subsequently, the gel is placed on a wick of filter paper that is immersed in a trough of buffer. A sheet of nitrocellulose membrane is placed on top of the gel and a large stack of absorbent paper (usually paper towels) on top of this. The buffer solution is drawn up by the absorbent paper, passes through the gel and nitrocellulose and carries the DNA out of the gel and onto the membrane. As before, the DNA is fixed in place by drying the filter in a vacuum. The net result is a replica on the nitrocellulose of the DNA fragment pattern from the agarose gel.

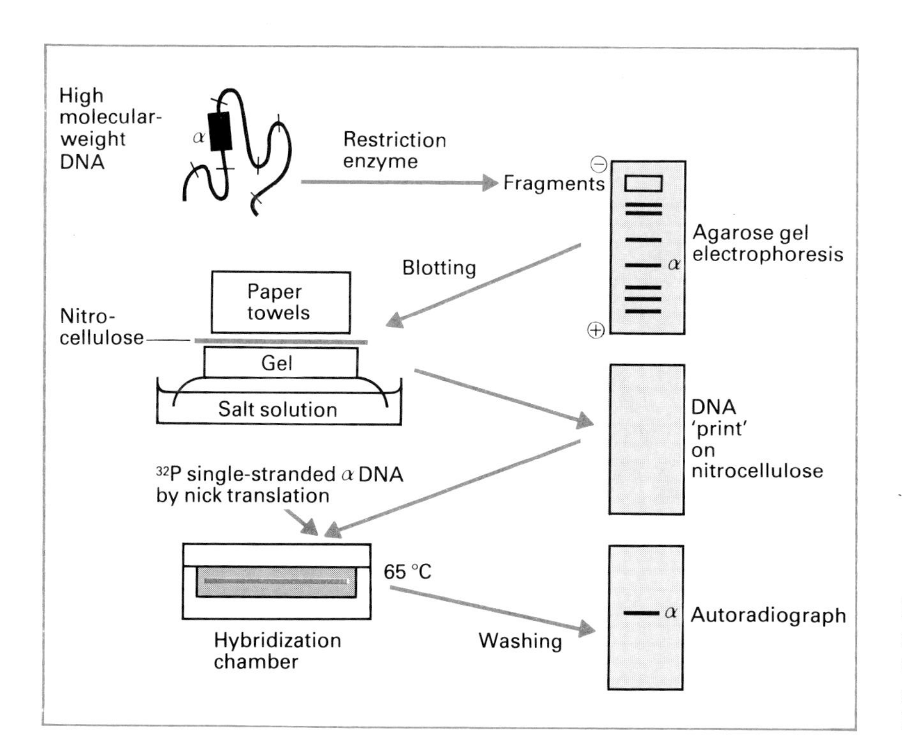

Fig. 3.6 Steps involved in the detection of an α-globin gene in a total genomic digest of human DNA using the technique of Southern blotting.

Example: Prenatal detection of inherited defects
Amniocentesis is a technique whereby a syringe is carefully inserted through the abdomen of a pregnant woman and used to withdraw a sample of the amniotic fluid which surrounds the fetus. This amniotic fluid contains some cells (amniocytes) derived from the developing child and, if desired, these can be cultured *in vitro* to increase their number. Since, with the exception of the antibody-producing B lymphocytes, the DNA in every somatic cell is identical, these amniotic cells are an accurate reflection of the genetic composition of the fetus. Using cloned gene probes it is possible to determine whether the fetus is carrying a particular inherited metabolic disorder and subsequently arrangements can be made for an abortion, if so desired.

The easiest form of gene defect to detect is where the loss of gene function is caused by a deletion of part or all of a gene sequence resulting in a reduction in size, or disappearance, of a particular DNA restriction fragment in the Southern blot. This has occurred in a number of thalassaemias (globin disorders) and certain types of osteogenesis imperfecta (a collagen disorder). Many other inherited diseases are not due to deletions but to single-base changes in a gene that result in a protein with an altered amino acid sequence. Sometimes this change of DNA sequence can be detected because it creates or destroys a site for a particular restriction enzyme. A good example of this is the sickle cell mutation where the change in DNA sequence destroys a site for the restriction enzyme *Mst* II leading to a different pattern of β-globin fragments on Southern blotting (Fig. 3.7).

Since this method of detecting *restriction fragment length polymorphisms* (RFLP) is limited by the sequence specificities of restriction enzymes, a more sophisticated method has been developed. In this approach two oligonucleotides, e.g. of 19 bases long, are chemically synthesized. One is perfectly complementary to the normal DNA sequence around the site of the mutation and the other is complementary to the mutant sequence. These two oligonucleotides, which differ in only one base out of 19, can be radiolabelled *in vitro* and used as probes in a Southern blot analysis of normal, carrier and patient DNA samples. By washing the blots at different temperatures after hybridization, conditions can be found in which the 'normal' probe will give a signal only with the normal gene sequence and not the mutant sequence, and vice versa. So far, this method has been used successfully to analyse the genes from patients with sickle-cell anaemia and deficiencies in α_1-antitrypsin (Fig. 3.8). This method should be applicable to any disease that is caused by a

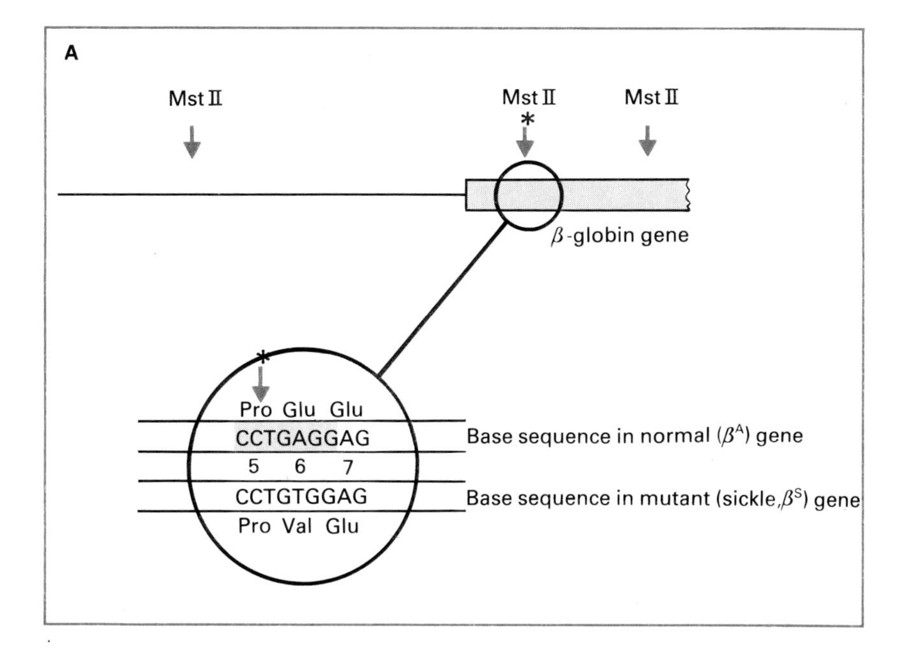

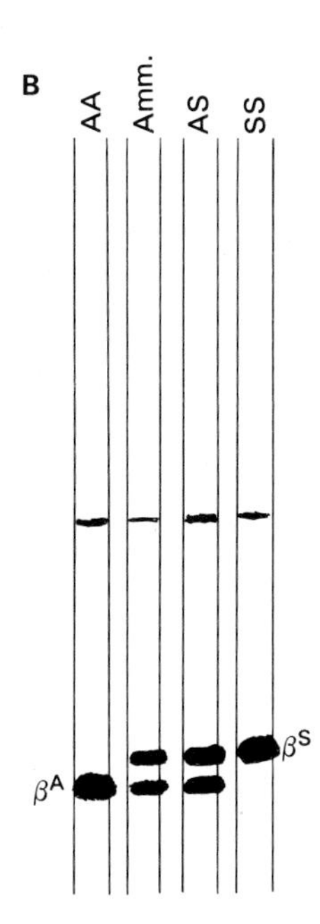

Fig. 3.7 Antenatal detection of sickle cell genes. Normal individuals are homozygous for the β^A allele, while sufferers from sickle-cell anaemia are homozygous for the β^s allele. Heterozygous individuals have the genotype $\beta^A\beta^s$. In sickle-cell anaemia, the 6th amino acid of β-globin is changed from glutamate to valine. (A) Location of recognition sequences for restriction endonuclease *Mst* II in and around the β-globin gene. The change of A → T in codon 6 of the β-globin gene destroys the recognition site (shaded) for *Mst* II as indicated by the asterisk. (B) Electrophoretic separation of *Mst* II-generated fragments of human control DNAs (AA, AS, SS) and DNA from amniocytes (Amn). After Southern blotting and probing with a cloned β-globin gene, the normal gene and the sickle gene can be clearly distinguished. Examination of the pattern for the amniocyte DNA indicates that the fetus has the genotype $\beta^A\beta^s$, i.e. it is heterozygous.

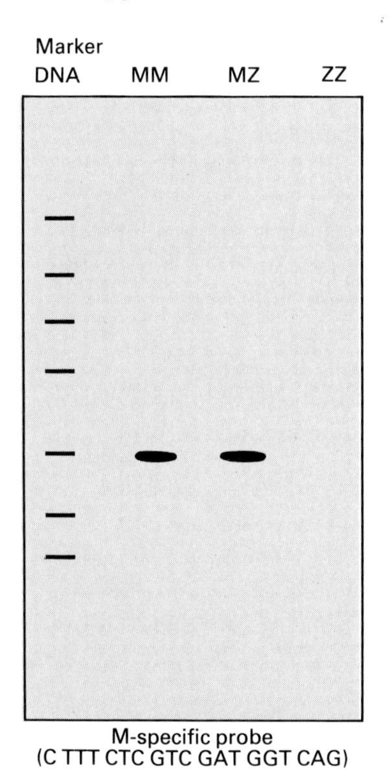

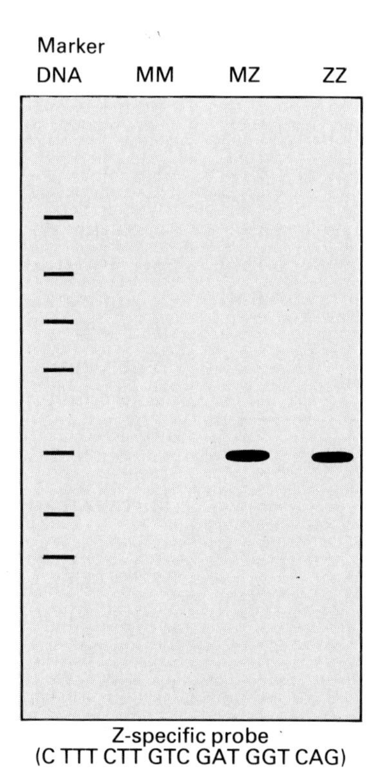

Fig. 3.8 Schematic representation of the use of oligonucleotide probes to detect the normal α_1-antitrypsin gene (M) and its Z variant. Human DNA obtained from normal (MM), heterozygous (MZ) and homozygous variant (ZZ) subjects is digested with a restriction endonuclease, electrophoresed and fragments Southern blotted onto a nitrocellulose membrane. The patterns shown were obtained on autoradiography of the filter following hybridization with either the normal (M-specific) or variant (Z-specific) probe.

common single-base mutation and is limited in sensitivity only by the ease of detection of the gene probe.

Example: Molecular fingerprinting It has long been the ambition of the forensic scientist to be able to identify the origin of blood and body-fluid stains with the same degree of certainty as fingerprints. By careful use of DNA probes and Southern blotting, Alec Jeffreys and his colleagues have provided the necessary method. The method can also be used to settle cases of doubtful paternity.

Dispersed throughout the human genome are simple, tandem-repetitive regions of DNA known as *minisatellites*. These minisatellites show substantial length polymorphism which probably arises from unequal genetic exchanges. The repeat elements in a particular group of human minisatellites share a common 10–15 base-pair core:

$$\text{GGAGGTGGGCAGG}\overset{\text{A}}{\text{G}}\text{G}.$$

A hybridization probe consisting of the core sequence repeated in tandem can detect many highly polymorphic minisatellites simultaneously. By using a series of variant core probes (Fig. 3.9), it is possible to detect a great number of hypervariable minisatellites thus producing DNA fingerprints which are completely specific to an individual (Fig. 3.10). Such fingerprints can be applied forensically (see below). Furthermore, suitable high molecular weight DNA can be isolated from such stains as blood and semen made on clothing several years previously and, of particular value in rape cases, sperm nuclei can be separated from the vaginal cellular debris present in semen-contaminated vaginal swabs.

Core sequence	$\text{GGAGGTGGGCAGG}\overset{\text{A}}{\text{G}}\text{G}$
Probe 33.6	$[(\text{AGGGCTGGAGG})_3]_{18}$
Probe 33.15	$(\text{AGAGGTGGGCAGGTGG})_{29}$
Probe 33.5	$(\text{GGG}\overset{\text{C}}{\text{A}}\text{GTGGGCAGGAGG})_{14}$

Fig. 3.9 Probes used for DNA fingerprinting.

The potential of DNA fingerprinting is best illustrated by the following example. A Ghanaian boy was initially refused entry into Britain because the immigration authorities were not satisfied that the woman claiming him as her son was in fact his mother. Analysis of serum proteins and erythrocyte antigens and enzymes showed that the alleged mother and son were related but could not determine whether the woman was the boy's mother or aunt. To complicate matters, the father was not available for analysis nor was the mother certain of the boy's paternity. DNA fingerprints from blood samples taken from the mother and three children who were undisputedly hers as well as the alleged son were prepared by Southern blot hybridization to two of the minisatellite probes shown in Fig. 3.9. Although the father was absent, most of his DNA fingerprint could be reconstructed from paternal-specific DNA fragments present in at least one of the three undisputed siblings but absent from the mother (Fig. 3.10). The DNA fingerprint of the alleged son contained 61 scorable fragments, all of which were present in the mother and/or at least one of the siblings. Analysis of the data showed the following.

1 The probability that either the mother or the father by chance possess all 61 of the alleged son's bands is 7×10^{22}. Clearly the alleged son is part of the family.

2 There were 25 maternal-specific fragments in the 61 identified in the alleged son and the chance probability of this is 2×10^{-15}. Thus the mother and alleged son are related.

3 If the alleged mother of the boy in question is in fact a maternal aunt, the chance of her sharing the 25 maternal-specific fragments with her sister is 6×10^{-6}.

When presented with this data, as well as results from conventional marker analysis, the immigration authorities allowed the boy residence in Britain.

Further reading

GENERAL

Dorkins H.R. & Davies K.E. (1985) Recombinant DNA technology in the clinical sciences. *Trends in Biotechnology* **3**, 195–9.

Highfield P.E. & Dougan G. (1985) DNA probes for microbial diagnosis. *Medical Laboratory Sciences* **42**, 352–60.

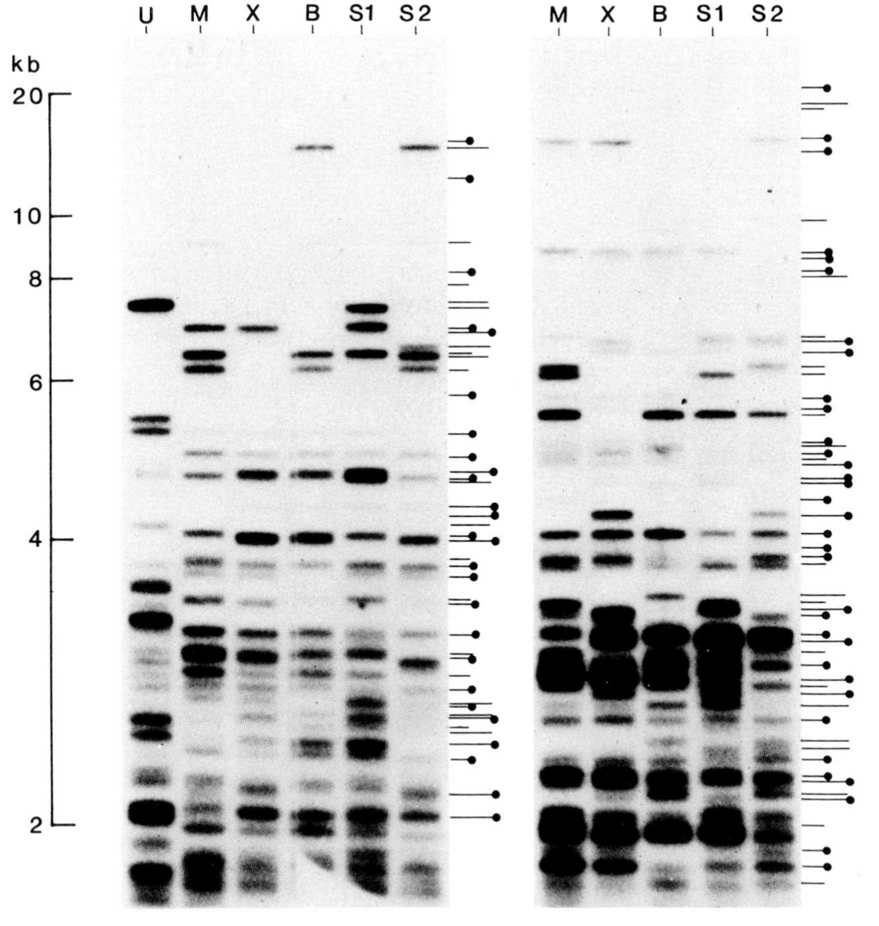

Fig. 3.10 DNA fingerprints of a Ghanaian family involved in an immigration dispute. Fingerprints of blood DNA are shown for the mother (M), the boy in dispute (X), his brother (B), sisters (S1, S2) and an unrelated individual (U). Fragments present in the mother's (M) DNA are indicated by a short horizontal line (to the right of each fingerprint); paternal fragments absent from M but present in at least one of the undisputed siblings (B, S1, S2) are marked with a long line. Maternal and paternal fragments transmitted to X are shown with a dot.

The lefthand fingerprint was obtained with probe 33.15 and the righthand fingerprint with probe 33.6. (Photo courtesy of Dr. A. Jeffreys and the editor of *Nature*.)

SPECIFIC

Beckmann J.S. & Bar-Joseph M. (1986) The use of synthetic DNA probes in breeders' rights protection: a proposal to superimpose an alpha-numerical code on the DNA. *Trends in Biotechnology* **4**, 230–2.

Cox D.W., Woo S.L.C. & Mansfield T. (1985) DNA restriction fragments associated with alpha$_1$-antitrypsin indicate a single origin for deficiency allele PIZ. *Nature* **316**, 79–81.

Gill P., Jeffreys A.J. & Werrett D.J. (1985) Forensic applications of DNA 'fingerprints'. *Nature* **318**, 577–9.

Jeffreys A.J., Brookfield J.F.Y. & Semeonoff R. (1985) Positive identification of an immigration test-case using human DNA fingerprints. *Nature* **317**, 818–19.

Law D.J., Frusard P.M. & Rucknagel D.L. (1984) Highly sensitive and rapid gene mapping using miniaturised blot hybridization; application to prenatal diagnosis. *Gene* **28**, 153–8.

Newmark P. (1986) Danger of delay for genetic tests. *Nature* **321**, 557.

Pace N.R., Stahl D.A., Lane D.J. & Olsen G.J. (1985) Analysing natural microbial populations by rRNA sequences. *American Society for Microbiology News* **51**, 4–12.

Saiki R.K., Scharf S., Faloona F., Mullis K.B., Horn G.T., Erlich H.A. & Arnheim N. (1985) Enzymatic amplification of β-globin genomic sequences and restriction site analysis for diagnosis of sickle cell anaemia. *Science* **230**, 1350–4.

Sensabaugh G.F. (1986) Forensic biology — is recombinant DNA technology in its future? *Journal of Forensic Sciences* **31**, 393–6.

4/Protein Engineering

INTRODUCTION

Many of the traditional fermentation processes for the manufacture of antibiotics, enzymes and amino acids do not utilize wild-type microorganisms because the yield of product is too low. Rather, the producing organism is subjected to successive rounds of mutagenesis until high-yielding strains are isolated. The improvement in yield is brought about by changes in proteins but what changes have been made, or even what proteins have been affected, is seldom known. As molecular biologists have learnt more about how mutations arise and how damage to DNA is repaired, the efficiency of random mutagenesis protocols has been improved. Similarly, as geneticists probing gene–protein relationships have devised clever selection procedures for rare mutants, their techniques and tricks have been adopted by those involved in strain selection. Despite the technical improvements which have been made the shotgun mutagenesis approach is limited in what it can achieve.

One of the spin-offs from recombinant DNA technology has been the development of techniques for *directed mutagenesis* of genes. There are two basic techniques. In *site-directed mutagenesis* an oligonucleotide consisting of a defined mutant sequence is hybridized to its complementary sequence in a clone of wild-type DNA. The oligonucleotide serves as a primer for in-vitro enzymatic DNA synthesis of regions that are to remain genotypically wild type. A double-stranded heteroduplex is formed, which subsequently is segregated *in vivo* into separate mutant and wild-type clones. In this way one or more changes can be introduced into a particular region of a gene. An alternative approach has been to use chemical synthesis of oligonucleotides to generate genes with specific mutations. In addition, a simple construct can be produced that allows for the replacement of segments by a number of mutated oligonucleotide 'cartridges' positioned between restriction endonuclease sites. The advantage of the gene synthesis approach is that it permits not only the introduction of point mutations but also the production of proteins in which entire polypeptide sequences have been precisely added, deleted or replaced.

This new-found ability to alter the amino acid sequence of a protein at will means that it is not unreasonable to consider improving proteins to meet a specific requirement (Table 4.1). This improvement is particularly relevant if it is planned to use an enzyme in an industrial process where the desired substrate or product is somewhat different from the physiological one or where the chemical conditions for the reaction are decidedly non-physiological. However, such *protein engineering* is much more complex than it might appear at first sight. Before rational changes can be made in a protein it is extremely useful to know its three-dimensional structure and what residues are important in the maintenance of this structure. If the protein has enzymic activity, it is important to know which amino acid residues have a catalytic function. This information is lacking for most proteins. Even where such information is available it is difficult to predict what effects specific changes will produce. Despite the fact that there is still a large degree of empiricism in protein engineering, a number of successes have been achieved.

Table 4.1 Potentially useful modifications to proteins.

Aspect	Change of interest
Stability	Increased thermostability
	Increased stability at low/high pH values
	Resistance to oxidative inactivation
Kinetics	Increased maximum velocity
	Increased affinity for substrate
	Altered substrate specificity
	Resistance to substrate/product inhibition
Biology	Altered spectrum of activity
	Altered substrate specificity

M13 vectors The vectors employed for site-directed mutagenesis are ones based on M13 and which were not described in Chapter 2. M13 is a rod-shaped bacteriophage which contains a single-stranded DNA genome within a coat of viral protein. When M13 infects a susceptible cell the single-stranded viral DNA, or plus strand as it is often known, is converted to a double-stranded replicative form (RF). Initially the RF is amplified in a semi-conservative mode but replication later becomes asymmetric such that only plus strands are synthesized (Fig. 4.1). These plus strands are packaged into phage particles from which they are readily isolated and, once isolated, are ideal for site-directed mutagenesis.

To convert M13 into a cloning vector it is necessary to have convenient restriction sites in a non-essential part of the genome. Although all the genes of M13 are essential for phage multiplication, there is a non-coding region (*intergenic space*) close to the origin of replication (Fig. 4.2). A series of DNA fragments carrying multiple restriction sites have been introduced into this space to generate the M13mp series of vectors. In practice the double-stranded RF form of the vector is isolated from infected cells and genes to be cloned are inserted in the usual way. Single-stranded versions of the cloned gene, ideal for site-directed mutagenesis, are obtained by extraction of the DNA from phage particles.

Basic site-directed mutagenesis To introduce a base change into a gene cloned in M13 a synthetic oligonucleotide, optimally 15–20 bases long, is synthesized. This oligonucleotide is complementary to the viral plus strand in the region to be mutated, except for that base which is to be changed (Fig. 4.3). The oligonucleotide is hybridized to the single-stranded template and acts as a primer for complementary strand synthesis by the Klenow fragment of DNA polymerase. The heteroduplex so formed is transformed into *Escherichia coli* where it undergoes replication and eventually gives rise to progeny phage, some of which will carry the desired mutant gene.

In theory, 50% of the progeny phage should carry the desired mutation but in practice the frequency may be much less. There are many reasons for this. For example, DNA repair mechanisms within the cell may correct the mismatches: G/T, A/C, G/G and A/A mismatches are known to be corrected with high efficiency. One way of minimizing mismatch repair is to transform the heteroduplex into cells lacking the appropriate enzymes, the so-called *mutator strains*.

Many different strategies have been developed to increase the frequency with which mutants are isolated. Of these, the most successful is that using thionucleotides to discriminate between the parental and mutant strands (Fig. 4.4). As before, the mutagenic oligonucleotide is annealed to the single-stranded template and extended by Klenow polymerase in the presence of DNA ligase to generate a mutant heteroduplex. If this step is carried out in the presence of a thionucleotide, the phosphorothioate DNA which is produced is resistant to cleavage with certain restriction enzymes. As a con-

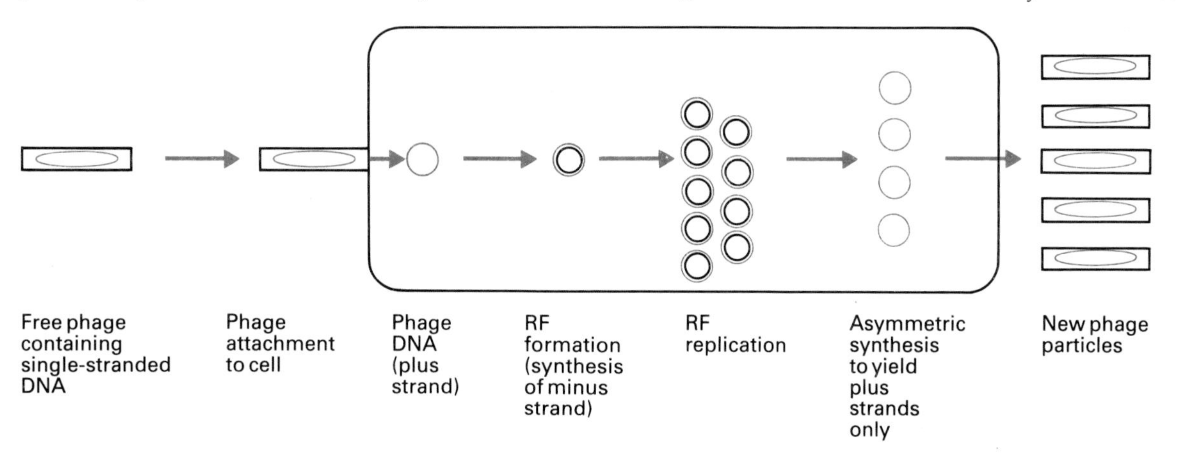

Fig. 4.1 Schematic representation of the life-cycle of bacteriophage M13, which contains a single-stranded circular DNA genome. The double-stranded DNA molecule is known as a replicative form (RF).

— ACC ATG ATT ACG CCA AGC TTG CAT GCC TGC AGG TCG ACT CTA GAG GAT CCC CGG GTA CCG AGC TCG AAT TCA CTG GCC —

Hind III, *Sph* I, *Pst* I, *Sal* I / *Acc* I / *Hinc* II, *Xba* I, *Bam* HI, *Xma* I / *Sma* I, *Kpn* I, *Sst* I, *Eco* RI, *Hae* III

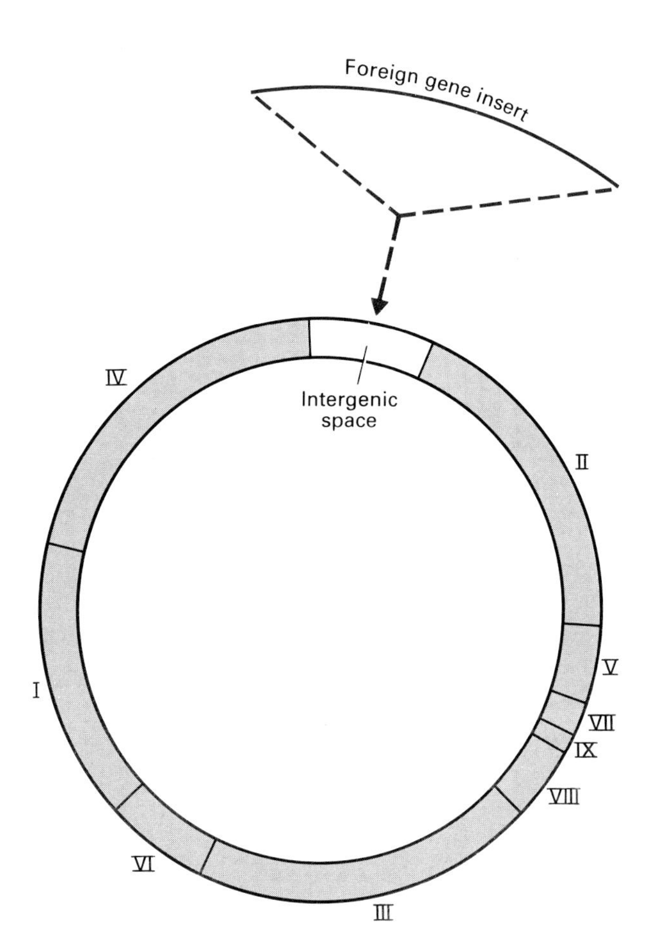

Fig. 4.2 The genetic map of bacteriophage M13 showing the location of the intergenic space used for gene cloning. The box shows a typical sequence which might be inserted.

sequence, single-strand nicks are generated in DNA containing one phosphorothioate and one non-phosphorothioate strand. Such nicks present sites for exonuclease attack permitting digestion of all or part of the non-mutant strand. The mutant strand which remains then is used as a template to reconstruct a double-stranded homoduplex mutant molecule *in vitro*.

Multiple mutations Site-directed mutagenesis need not be restricted to single base changes. By appropriate choice of oligonucleotides it is possible to introduce multiple-point mutations, insertions and deletions (Fig. 4.5). Thus by using a 30 base oligonucleotide (referred to as a 30-mer, meaning a polymer of 30 nucleotides) 11 base changes were introduced into a single region of a gene. Multiple base changes at the *same* position can be made in a single experiment by using mixed oligonucleotide primers. Large deletions can be made between two sites if one half of the primer is complementary to the first site and the other half complementary to the second site. In this way an 18-mer has been used to delete over 1100 nucleotides. Insertions can be made likewise, although the length of the insert

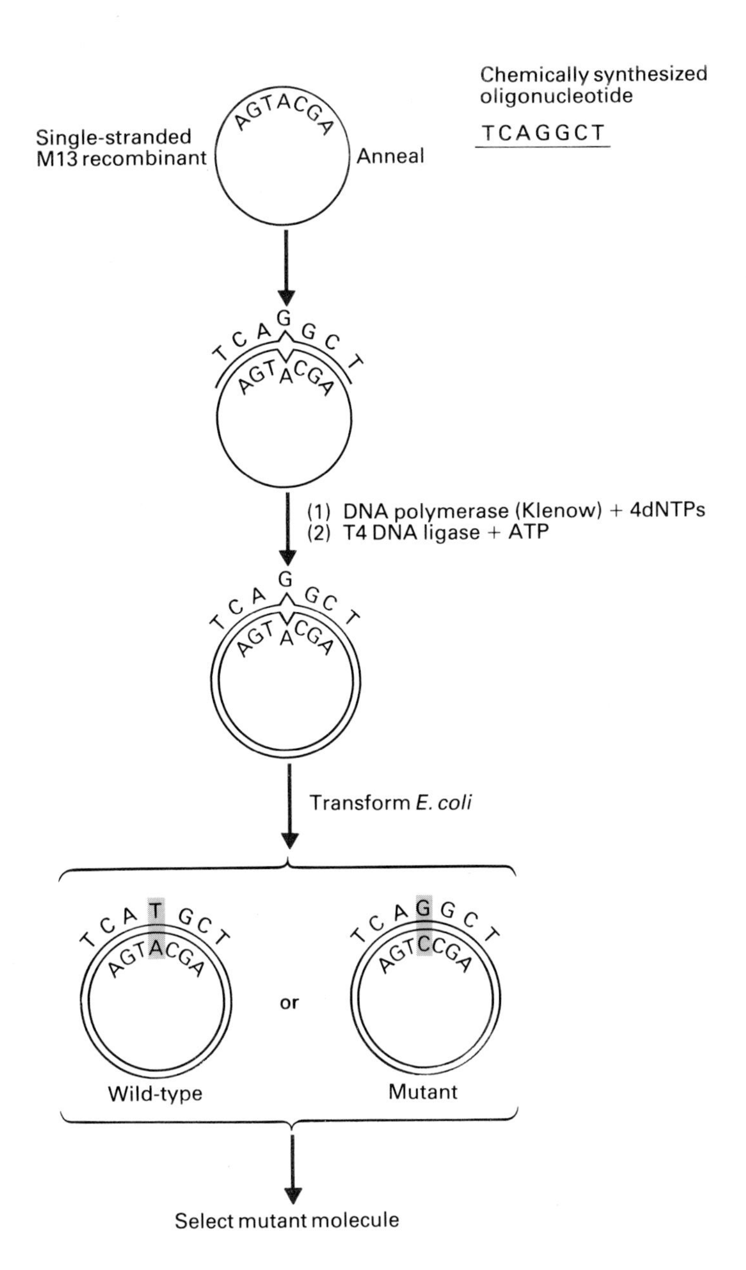

Fig. 4.3 The principle of oligonucleotide-directed mutagenesis.

is limited by the length of the primer. Using normal-length oligonucleotides (e.g. less than 40-mers) the largest insert which can be achieved is 10–15 bases.

INCREASED PROTEIN STABILITY

The goal of many protein engineering programmes is to achieve increased protein stability. This may be either increased half-life or thermostability of an enzyme used in an enzyme reactor, increased shelf-life of a protein for therapeutic use or resistance to inactivation by oxidation of important amino acid residues. These goals have been achieved as the examples given below show. Increased thermostability is particularly important since there are benefits from operating enzyme reactors at elevated temperatures; for example, reaction times are shorter and there is less risk of microbial spoilage of the enzyme. Furthermore, cooling of reactors is not always efficient, particularly when exothermic reactions are being performed.

Example: Increased thermostability of lysozyme One theoretical way of increasing the thermal stability of a protein is to introduce a disulphide bond where none already exists in the molecule. Although not of

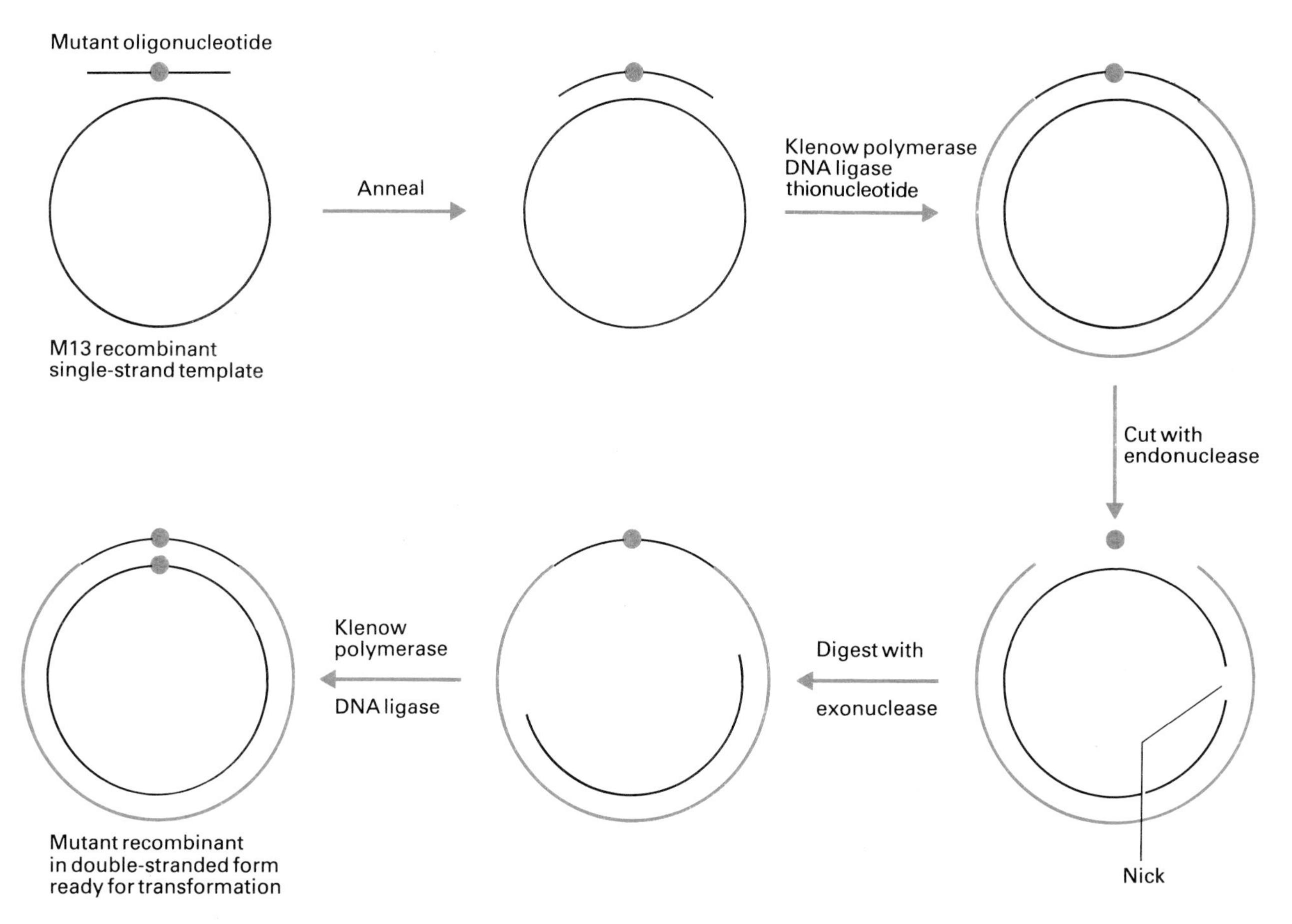

Fig. 4.4 The principle of thionucleotide selection of mutant DNA molecules (see text for details). The mutant base is shown as a red spot. DNA synthesized in the presence of thionucleotides is shown by the heavy red line.

commercial importance, T4 lysozyme is the only enzyme in which this has been achieved. In T4 lysozyme there are two cysteine residues (54 and 97) which are unpaired. Careful examination of the three-dimensional structure indicated that conversion of isoleucine residue 3 to a cysteine should enable a disulphide bridge to form with cysteine 97. Site-directed mutagenesis was used to convert the ATA isoleucine codon in a cloned T4 lysozyme gene to a TGT cysteine codon and biochemical studies on the mutant enzyme confirmed that a cross-link did form as expected. Under oxidizing conditions the mutant and wild-type enzymes had the same enzymic activity but the mutant had greater thermostability. At 67°C the half-life of the wild-type enzyme was 11 min, whereas that of the mutant enzyme was 28 min. Furthermore, the activity of the mutant enzyme never fell below 50% of maximum, whereas the wild-type enzyme's activity dropped to 0.2% after 180 min.

Example: Enhanced specific activity of recombinant interferon-β A different kind of stability has been achieved with human interferon-β, a protein with 3 cysteine residues. When interferon-β is synthesized in *E. coli* it forms aggregates and the specific activity of the protein is 10^6 antiviral units/mg, one hundred times less than fibroblast-derived material. One possible explanation for the low specific activity of the recombinant material is incorrect disulphide bond formation. Thus when cysteine residue 17, which should be present as a free thiol, was changed to serine the specific activity of the mutant interferon was 10^8 units/mg, i.e. it had full specific activity. An additional benefit of this cysteine to serine change was increased stability on prolonged storage, an important feature of any protein which is to be sold for therapeutic use.

Example: Improved α_1 antitrypsin α_1-antitrypsin (AAT) is the archetypal member of a family of

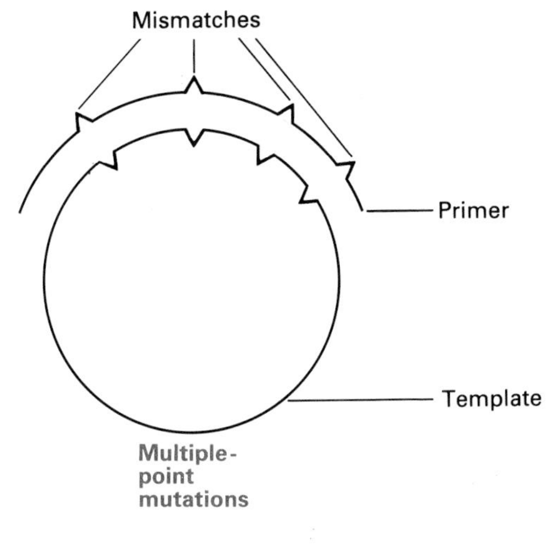

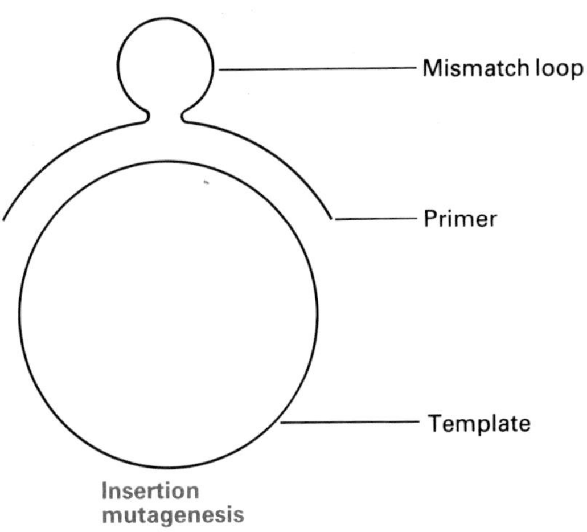

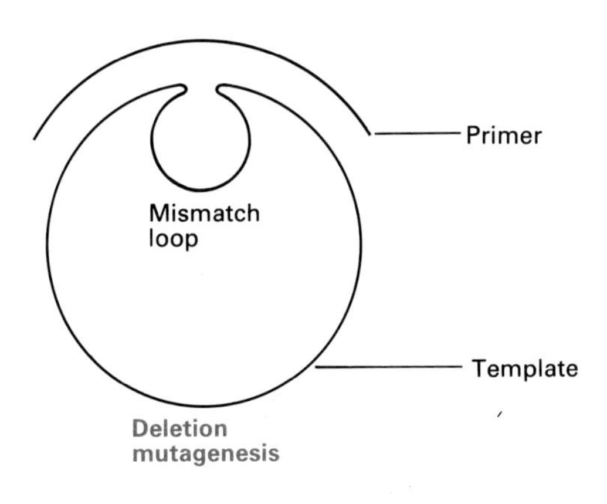

Fig. 4.5 The use of oligonucleotide-directed mutagenesis to generate multiple-point mutations, insertions and deletions.

homologous proteins known as serine protease inhibitors or serpins. *In vivo* the function of AAT is the inhibition of neutrophil elastase. Inhibition occurs by a rapid association between the two proteins followed by cleavage of the AAT between methionine residue 358 and serine residue 359 (Fig. 4.6). After cleavage there is negligible dissociation of the complex.

Oxidative stress, as occurs in a number of pathological conditions (see Box), results in the conver-

α_1-antitrypsin, emphysema and smoking

Cumulative damage to lung tissue by neutrophil elastase is thought to be responsible for the development of emphysema, an irreversible lung disease characterized by loss of lung elasticity. The primary defence against elastase damage is α_1-antitrypsin (AAT), a glycosylated serum protein of 394 amino acids. The function of AAT is known because its genetic deficiency leads to a premature breakdown of connective tissue. In healthy individuals the AAT diffuses into tissue spaces where it complexes with neutrophil elastase and these complexes are subsequently removed from the circulation and catabolized in the liver and spleen.

Smokers are more prone to emphysema because smoking results in an increased concentration of leucocytes in the lung and consequently increased exposure to neutrophil elastase. In addition, leucocytes liberate oxygen-free radicals and these can oxidize the critical methionine residue 358 (see main text) rendering the AAT ineffective. Exposure of AAT to oxidants can also occur with ozone pollution or therapeutic use of pure oxygen.

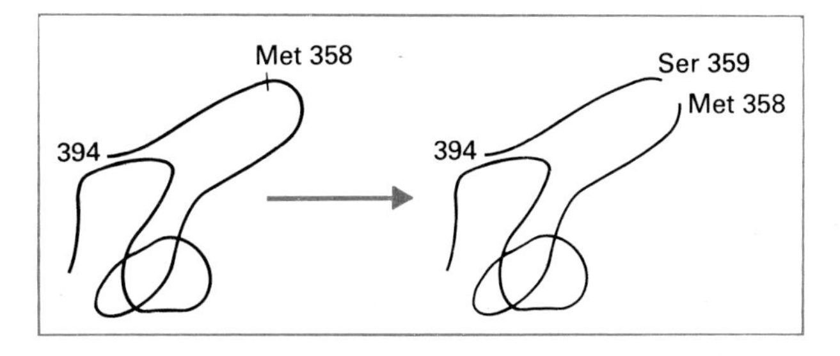

Fig. 4.6 The cleavage of α_1-antitrypsin on binding to neutrophil elastase.

sion of methionine 358 to methionine sulphoxide. Since methionine sulphoxide is much bulkier than methionine it does not fit into the active site of elastase and hence oxidized AAT is a poor inhibitor. By means of site-directed mutagenesis an oxidation-resistant mutant of AAT has been constructed by replacing methionine 358 with valine. In a laboratory model of inflammation, the degradation of basement membrane collagen by stimulated neutrophils, the modified AAT was an effective inhibitor of elastase and was not inactivated by oxidation. Clinically this could be important since intravenous replacement therapy with plasma concentrates of AAT is already being tested on patients with a genetic deficiency in AAT production.

ALTERING THE KINETIC PROPERTIES OF ENZYMES

Many enzymes are used in large-scale industrial processes and a number of examples are given in later chapters (see Tables 6.8 and 7.3). Effective use of these enzymes requires a knowledge of enzyme kinetics (see Box). For most applications the enzymes are used under conditions where there is a vast excess of substrate and v_{max} is the only kinetic parameter of importance. It should be noted that

Basic enzyme kinetics

Most enzymatic reactions use more than one substrate in concert but it is easier to explain the basic principles of enzyme kinetics by considering a one-substrate reaction. Such a reaction can be summarized as:

Enzyme (E) + substrate (S) ⇄ enzyme–substrate complex (ES) ⇄ enzyme + products (P).

From this it should be apparent that increasing the substrate concentration at a fixed level of enzyme increases the reaction velocity (Diagram 1). Maximum velocity (v_{max}) is achieved when all the enzyme is in the form of the ES complex.

For the purposes of kinetic analysis it is convenient to examine that period of the enzymatic reaction known as the *steady state*. This occurs shortly after mixing enzyme and

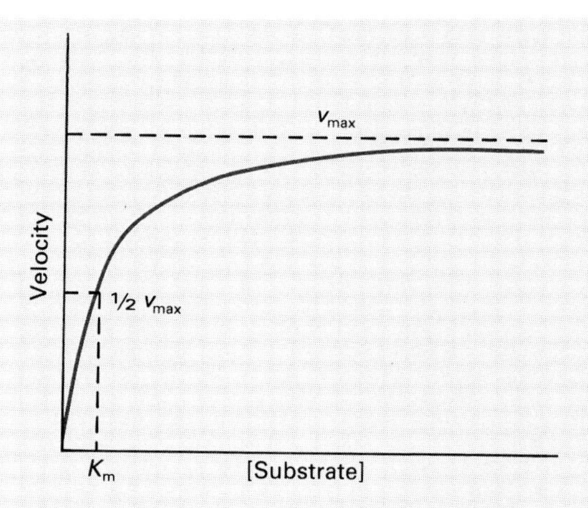

Diagram 1 Effect of substrate concentration on the rate of an enzyme-catalysed reaction.

substrate and constitutes the time interval when the rate of reaction is a linear function of time (Diagram 2). Usually substrate is present in much higher molar concentrations than enzyme and, since the initial period of the reaction is being examined, the free substrate concentration is approximately equal to the total substrate added to the reaction mixture.

There are four rate constants associated with the above reaction sequence:

$$E + S \underset{k_3}{\overset{k_1}{\rightleftharpoons}} ES \underset{k_4}{\overset{k_2}{\rightleftharpoons}} E + P.$$

If only the *initial* rate of reaction is being

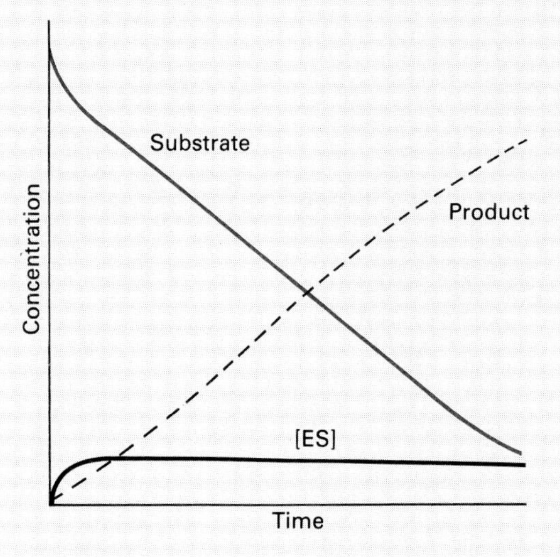

Diagram 2 Time-course of the formation of product and enzyme-substrate complex in an enzyme-catalysed reaction.

considered, the rate constant k_4 can be ignored because not enough product will be present to make the reverse reaction proceed at a significant rate. If v is the rate of product formation then

$$v = k_2[\mathrm{E}]_t \qquad (1)$$

and since the maximum initial rate v_{max} should be obtained when all the enzyme is in the form of the ES complex,

$$v_{max} = k_2[\mathrm{E}]_t \qquad (2)$$

where $[\mathrm{E}]_t$ represents the total enzyme concentration, i.e. the sum of the free and combined enzyme. The kinetic constant k_2 is referred to as the *turnover number* and sometimes is symbolized by K_{cat}. From equation (2) it can be seen that the turnover number is equal to the number of substrate molecules converted into product in a given time (usually 1 s) by a mole of enzyme when the enzyme is fully complexed with substrate. For most enzymes this number falls in the range 10^1-10^4/s.

Under steady-state conditions the rate of formation of the ES complex is equal to the rate of its breakdown. This rate is known as K_m or the *Michaelis constant* and is represented by the expression:

$$K_m = \frac{k_2 + k_3}{k_1}. \qquad (3)$$

Numerically K_m is equal to the concentration (expressed in mol/l) of the substrate which gives half the maximum velocity. For any given enzyme the value of K_m depends on the substrate and the conditions of the reaction such as pH, temperature, etc. For most enzymes the value of K_m lies between 10^{-1} and 10^{-6} mol/l. The lower the value of K_m, the stronger the complex between enzyme and substrate.

v_{max} is usually measured during the initial stages of the reaction. As the reaction proceeds the concentration of P will increase and the rate constant k_4 (see Box) can no longer be ignored unless the reaction is irreversible. High concentrations of product can also inhibit the enzyme, leading to a decrease in the reaction velocity.

If an enzyme is being used to *destroy* a substrate present at low concentrations, the K_m of the enzyme is the key parameter. The best example is provided by asparaginase. In certain leukaemias the growth of the malignant white blood cells is dependent upon the presence in the bloodstream of asparagine. Thus intravenous administration of the enzyme asparaginase is therapeutically beneficial since it cleaves asparagine to aspartate. During a search for sources of asparaginase the puzzling discovery was made that not all asparaginases are effective in suppressing leukaemia. The explanation was eventually provided by kinetic analysis; the enzymes from the different sources differ widely in their K_m for asparagine. The administration of asparaginase will be therapeutically useful only if its K_m value is low enough to hydrolyse asparagine rapidly at the low concentration at which it is present in blood.

Example: Altering the K_{cat} *and* K_m *of subtilisin* The industrial applications of many enzymes are restricted because of oxidative inactivation in an analogous manner to that of AAT. This is especially true for enzymes containing methionine, cysteine or tryptophan residues in or around the active site. Like AAT, the protease subtilisin has a methionine residue in its active site and, being a component of biological detergents, subtilisin is exposed to oxidative stress during use! Using site-directed mutagenesis, variant enzymes have been obtained in which this methionine residue is replaced by any one of the other 19 natural amino acids. The different mutant enzymes were found to vary widely in activity (see Table 4.2 for representative examples). With the exception of the methionine to cysteine replacement, all had decreased activity. Mutants

Table 4.2 Kinetic constants for selected mutants of subtilisin in which methionine residue 222 is replaced by other amino acids.

Amino acid 222	K_{cat} (per s)	K_m (mol/l)
Methionine	50	1.4×10^{-4}
Cysteine	84	4.8×10^{-4}
Serine	27	6.3×10^{-4}
Alanine	40	7.3×10^{-4}
Leucine	5	2.6×10^{-4}

containing non-oxidizable amino acids, e.g. serine, alanine or leucine, were resistant to inactivation by 1 mol/l hydrogen peroxide, whereas the methionine and cysteine enzymes were rapidly inactivated (Fig. 4.7).

Evaluating this kinetic information in terms of the usefulness of the subtilisin mutants for inclusion in detergents is a little complex. With the exception of the cysteine replacement, all the mutants have a decreased turnover number and, at first sight, would appear to be less useful but that is in the absence of oxidative stress. If oxidative stress during use approximates to 1 mol/l hydrogen peroxide, then clearly the alanine-containing variant would be most useful (Fig. 4.7). If the stress approximates to 0.1 mol/l hydrogen peroxide, then perhaps the cysteine-containing variant would be more appropriate. However, there is another consideration. Unlike many enzymes of commercial importance, subtilisin in detergents operates at low substrate concentration. Thus the affinity of the enzyme for its substrate, i.e. the K_m value, cannot be ignored and the replacement of methionine by other amino acids leads to a decreased affinity.

Example: Altering the pH dependence of enzyme catalysis One prospect for protein engineering is the tailoring of the pH dependence of enzyme catalysis to optimize activity in industrial processes. Chemical studies have shown that the pH dependence of catalysis by serine proteases alters with changes in overall surface charge. Thus a possible way of modifying pH dependence would be to alter the electrostatic environment of the active site by protein engineering and so change the pK_a values of ionic catalytic groups. That such alterations are possible has been shown again with subtilisin where a single amino acid change, aspartate to serine, reduced the pH optimum by 0.3 units.

PROTEIN STRUCTURE DETERMINATION

Successful protein engineering of the kind discussed above depends upon the availability of reliable structural information. At present X-ray diffraction methods are the only techniques that can provide the required data, although two-dimensional nuclear magnetic resonance (NMR) may be an alternative procedure in the future. Protein crystallography is a very laborious process despite major advances in the collection and analysis of diffraction data. The most unpredictable step, and hence the one which is rate-limiting, is the preparation of protein crystals of the correct quality. Indeed, some workers have been forced to make use of the zero gravity aboard space shuttles to eliminate convective effects and improve crystallization!

The use of sophisticated computer graphics has eliminated the need for building physical models. The protein structure is simply fitted to the electron density map provided by X-ray diffraction analysis. In addition, the use of computers facilitates the estimation of the perturbations that would result from specific amino acid substitutions of a known protein structure.

MACRO-MODIFICATIONS TO PROTEINS

With site-directed mutagenesis the aim is to replace, via base changes in a gene, one or, at most, a

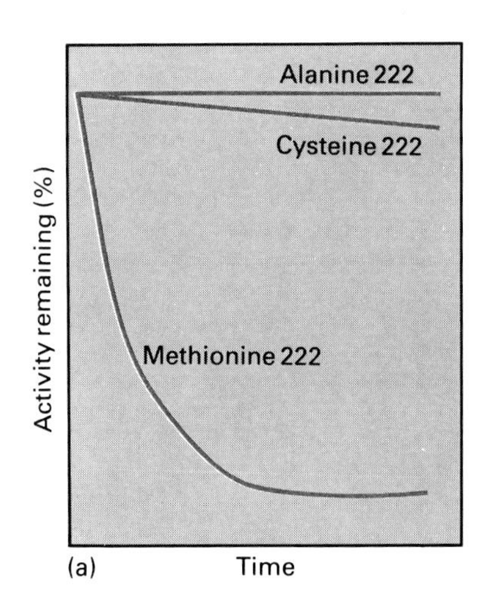

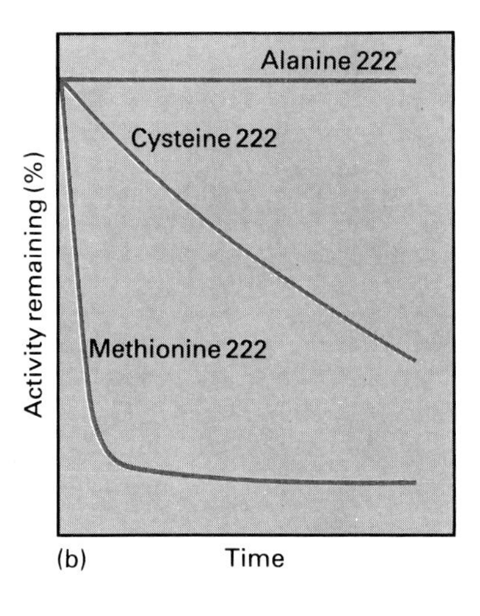

Fig. 4.7 Resistance to oxidation by subtilisin mutants in which methionine residue 222 is replaced by alanine or cysteine. Activity remaining after exposure to 0.1 mol/l (a) or 1.0 mol/l (b) hydrogen peroxide.

few amino acids in a protein sequence which will in some way 'improve' the protein in question. Much larger modifications can be made to a protein; for example, part of a gene can be deleted by eliminating a restriction fragment or by chemically synthesizing only part of a gene. In this way it is possible to produce the Klenow fragment of DNA polymerase (see Box, p. 24) free of contamination with the $3' \rightarrow 5'$ exonuclease activity associated with the intact enzyme. Conversely, it is possible to insert additional amino acids into a protein sequence and this has been done to create purification fusions (see below) and to improve the stability of foreign proteins synthesized in *E. coli*. Finally, all or part of one gene can be fused with all or part of another to generate completely novel proteins as has been done with the interferons (see below).

Example: Novel interferons Human interferon-α and -β elicit a number of biological responses such as conferring resistance to viral infections, macrophage and killer cell activation, inhibition of cell proliferation and modulation of the immune response. When administered clinically a reversible dose-limited toxicity is frequently observed and this could be because of the action of interferons on so many cell types. By constructing hybrids between different interferons it may be possible to produce novel interferons with superior clinical properties, e.g. increased potency or decreased toxicity. That this is feasible has been shown by the synthesis of a 'consensus' interferon-α with a 15-fold higher specific activity than any natural interferon-α. Hybrids also have been constructed in which portions of the interferon-β molecule have been replaced with corresponding segments from interferon-α. Some of these hybrid interferons have an altered antiviral or antiproliferative profile (Table 4.3).

Table 4.3 Properties of some modified interferons.

Amino acid change	Antiviral activity relative to interferon-β	Antiproliferative activity relative to interferon-β
Interferon-β residues 2–7 replaced by interferon-α_2 residues 1–5 and cysteine 17 replaced by serine	2.3	4.4
Interferon-β residues 9–56 replaced by interferon-α_1 residues 7–54	0.005	1.9
Interferon-β residues 82–105 replaced by interferon-α_1 residues 80–103	0.76	3.3

Genetically engineered antibodies have been produced in a similar way; for example, mouse and rat antibodies have been converted into ones with far greater similarity to human antibodies. This is discussed in more detail in Chapter 10 (see p. 125).

Example: Purification fusions Recombinant DNA technology can be used to minimize or eliminate some of the problems associated with protein purification, particularly when the protein is destined for therapeutic use. Many foreign proteins when produced in *E. coli* are particularly susceptible to protease attack such that little intact protein can be isolated. These labile proteins can be protected by producing them as fusions with *E. coli* proteins and such fusion proteins can facilitate purification. One generalized method which has been developed is to build a gene which encodes the target protein joined to β-galactosidase via a collagen peptide linker (Fig. 4.8). The tripartite protein first is purified by exploiting its affinity for resin-bound p-aminophenyl thiogalactoside residues and the target protein released by controlled digestion with collagenase. Any protein for which there is a suitable affinity purification method can be used instead of β-galactosidase, e.g. protein A-containing fusions can be purified on columns of immobilized immunoglobulin.

An alternative approach to purification fusions is polyarginine-tailing. In this method the gene sequence encoding the desired protein is extended by the inclusion at the 3' end of a number of codons for arginine. When such genes are expressed the resultant proteins have a polyarginine 'tail' which makes them more basic. Upon ion exchange chromatography such proteins are separated from the bulk of the host cell proteins which are more acidic (Fig. 4.9). The polyarginine tail is then removed with the enzyme carboxypeptidase B which for convenience can be immobilized. The de-tailed protein is rechromatographed on an identical ion-exchange resin to separate it from any remaining contaminating proteins which will be more basic.

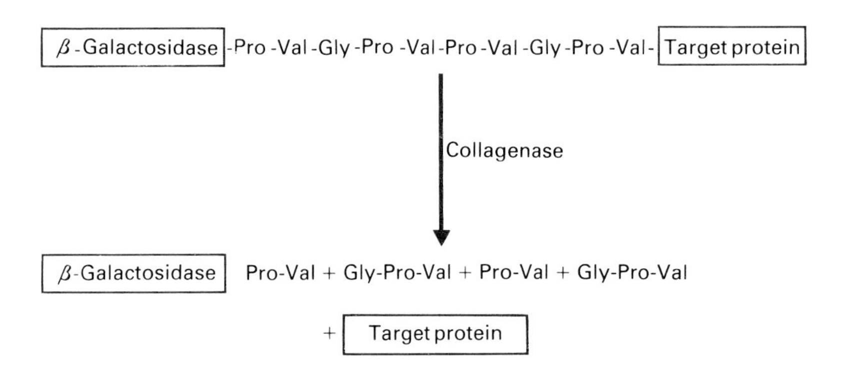

Fig. 4.8 A tripartite protein fusion which can be used to facilitate purification of a protein.

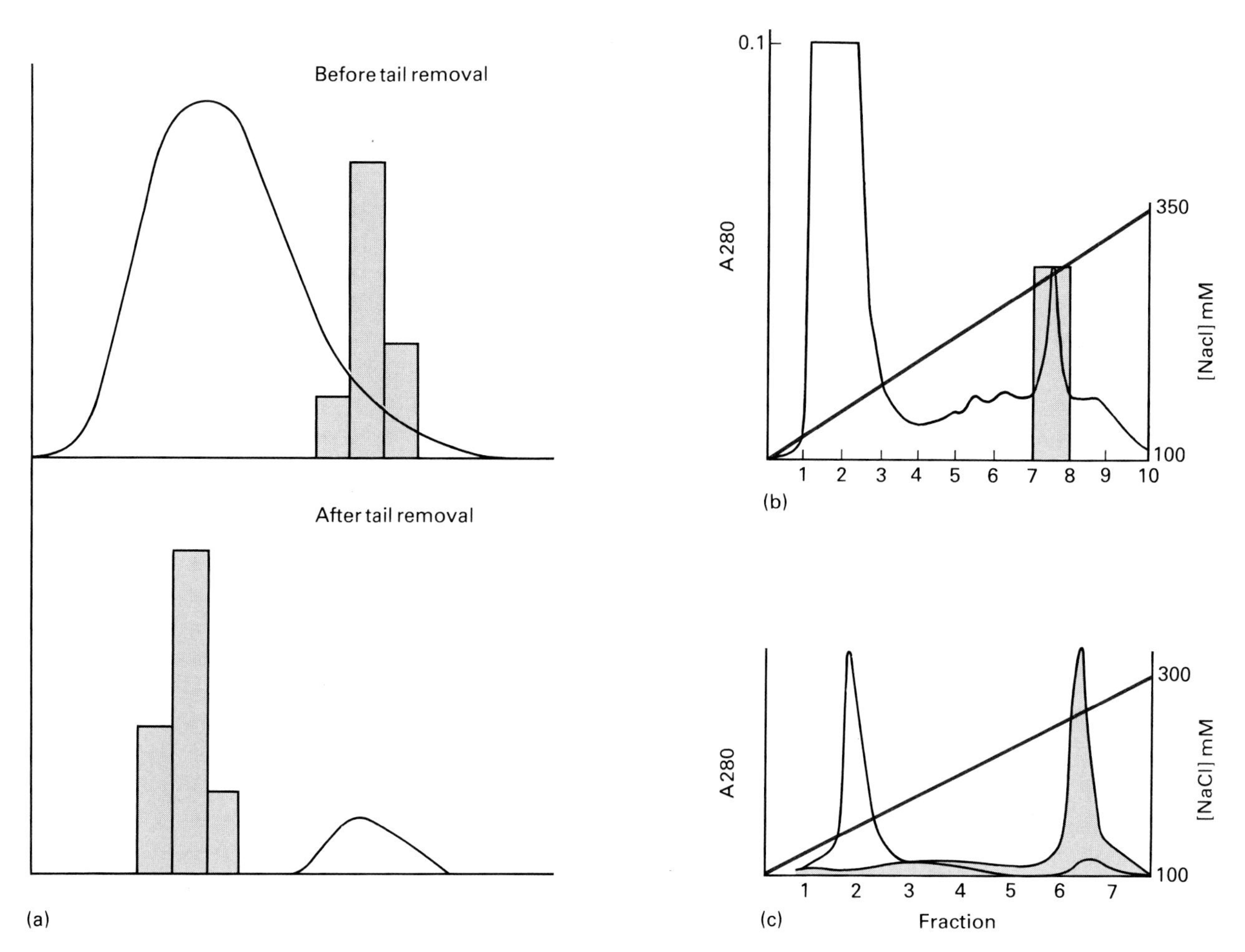

Fig. 4.9 The use of polyarginine-tailing to facilitate protein purification. (a) (*left*) Schematic representation of a hypothetical protein before and after enzymic removal of the C-terminal arginine residues. (b) (*top right*) Separation of polyarginine-tailed urogastrone from the bulk of the proteins in an *E. coli* cell extract. The red line shows the salt gradient and the tinted bar the urogastrone activity. (c) (*bottom right*) Chromatographic behaviour of tailed and untailed urogastrone. The red line shows the salt gradient.

Further reading

GENERAL

Baum R.M. (1986) Enzyme chemistry set to advance as new techniques are applied. *Chemical and Engineering News* **65**, 7–14.

Blundell T. & Sternberg M.J.E. (1985) Computer aided design in protein engineering. *Trends in Biotechnology* **3**, 228–35.

Botstein D. & Shortle D. (1985) Strategies and applications of *in vitro* mutagenesis. *Science* **229**, 1193–201.

Kabsch W. & Rosch P. (1986) Protein structure determination. *Nature* **321**, 469–70.

Oi V.T. & Morrison S.L. (1986) Chimeric antibodies. *Biotechniques* **4**, 214–21.

Primrose S.B. (1986) The application of genetically engineered microorganisms in the production of drugs. *Journal of Applied Bacteriology* **61**, 99–116.

Robson B. & Garnier J. (1986) *Introduction to Proteins and Protein Engineering*. Elsevier, Amsterdam.

Ulmer K.M. (1983) Protein engineering. *Science* **219**, 666–71.

SPECIFIC

Carrell R. & Travis J. (1985) Alpha$_1$-antitrypsin and the serpins: variation and countervariation. *Trends in Biochemical Sciences* **10**, 20–4.

Estell D.A., Graycar T.P. & Wells J.A. (1985) Engineering an enzyme by site-directed mutagenesis to be resistant to chemical oxidation. *Journal of Biological Chemistry* **260**, 6518–21.

Fersht A.R., Leatherbarrow R.J. & Wells T.N.C. (1986) Binding energy and catalysis: a lesson from protein engineering of the tyrosyl-tRNA synthetase. *Trends in Biochemical Sciences* **11**, 321–5.

Fersht A.R., Shi J-P., Knill-Jones J., Lowe D.M., Wilkinson A.J., Blow D.M., Brick P., Carter P., Waye M.M.Y. & Winter G. (1985) Hydrogen bonding and biological specificity analysed by protein engineering. *Nature* **314**, 235–8.

Imanaka T., Shibazaki M. & Takagi M. (1986) A new way of enhancing the thermostability of proteases. *Nature* **324**, 695–7.

Kempe T., Chow F., Peterson S.M., Baker P., Hays W., Opperman G., L'Italien J.J., Long G. & Paulson B. (1986) Production and characterization of growth hormone releasing factor analogs through recombinant DNA and chemical techniques. *Bio/technology* **4**, 565–8.

Taylor J.W., Schmidt W., Cosstick R., Okruszek A. & Eckstein F. (1985) The use of phosphorothioate-modified DNA in restriction enzyme reactions to prepare nicked DNA. *Nucleic Acids Research* **13**, 8749–64.

Part III
The Exploitation of Microbes

5/The Large-scale Cultivation of Microbes

INTRODUCTION

Much of biotechnology centres around the large-scale cultivation (10–500 000 l) of microbes for the production of cells themselves, e.g. biomass, or the production of useful metabolites or proteins. Such large-scale cultures are often referred to as industrial fermentations. *Sensu strictu*, a fermentation is a biological process occurring in the absence of air (oxygen). The first large-scale microbial processes to be developed, brewing and acetone-butanol production, are truly fermentations. Ethanol is formed by yeast cells as a means of regenerating oxidized pyridine nucleotides in the absence of oxygen (Fig. 5.1). Acetone and butanol are formed for the same reason but in this instance the producing organism is a strict anaerobe. However the term 'industrial fermentation' is now applied to any large-scale cultivation of microbes even though most of them are aerobic. This is ironic for oxygen supply is the single most important factor limiting the efficiency

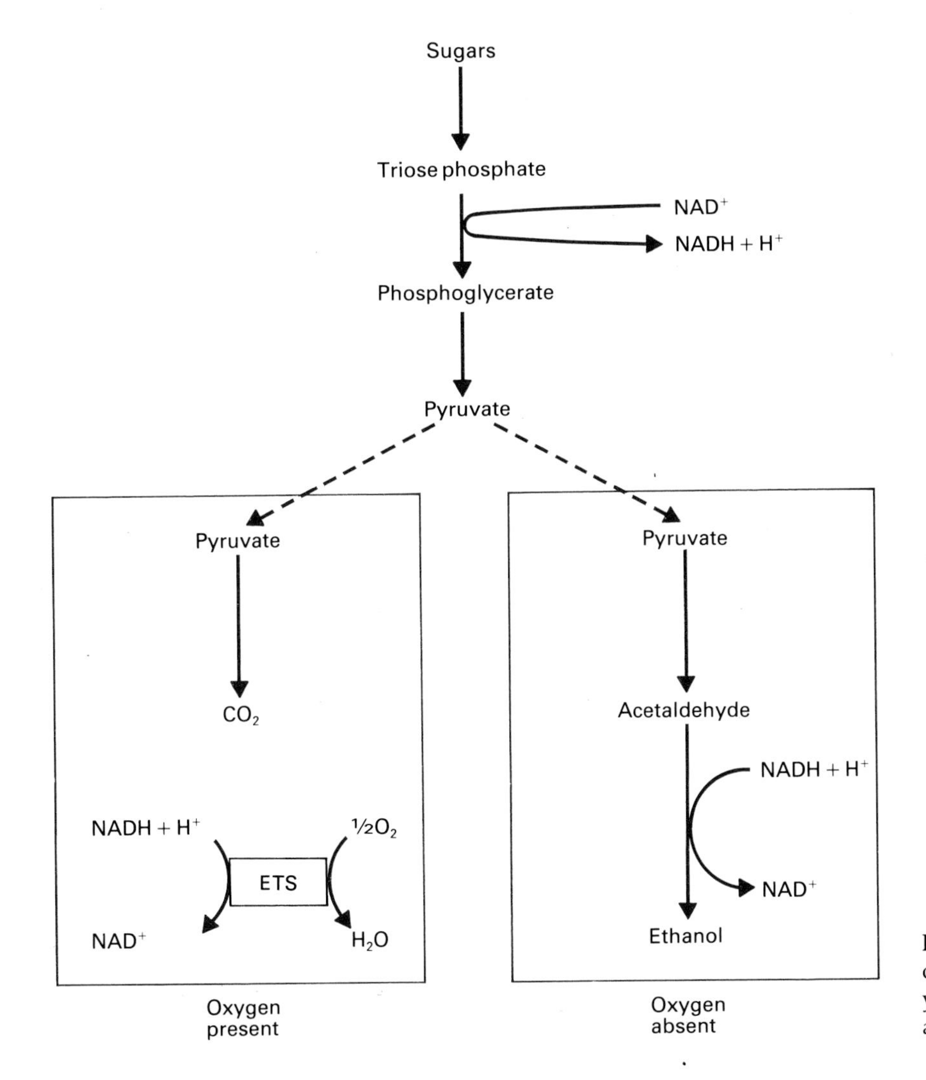

Fig. 5.1 The regeneration of oxidized pyridine nucleotides by yeast cells in the presence and absence of oxygen.

of aerobic processes and modern fermenter design and practice centres around the provision of adequate amounts of oxygen! Obligate aerobes require oxygen for growth: limit the supply of oxygen and growth ceases. Facultative anaerobes can grow in the absence of oxygen but growth rates are lower, leading to longer process times, and cell yields are greatly reduced with a concomitant reduction in product levels.

THE PROBLEM OF OXYGEN SUPPLY

A rapidly growing culture has a very high demand for oxygen. More correctly, it has a high demand for dissolved oxygen. The gas must be dissolved in the growth medium so that it can interact with the membrane-bound electron transport system to effect the oxidation of reduced pyridine nucleotide cofactors (Fig. 5.2). The fundamental problem in supplying sufficient oxygen is that oxygen is very insoluble in aqueous systems (Table 5.1). It should be noted from Table 5.1 that the solubility of oxygen decreases as the temperature increases. The solubility also decreases as the concentration of dissolved solutes increases. However, the solubility of oxygen in the growth medium can be increased by an increase in the oxygen pressure of the gas phase. A second problem is that in many fermentations, particularly those involving growth of filamentous organisms such as fungi or actinomycetes, the culture broth is extremely viscous and the controlling factor is transporting dissolved oxygen from the bulk liquid to the cell and not from the gas phase to the liquid phase.

For small-scale culture, e.g. less than 1 l, it is possible to supply adequate amounts of oxygen by growing the microbe in an Erlenmeyer flask which is agitated constantly. Agitation facilitates the transfer of oxygen from the gas phase in the flask to the liquid phase. Provided that the volume of growth medium does not exceed 10–20% of the flask volume it is possible to achieve cell densities of 1–2 g dry wt/l before oxygen becomes limiting. Once the cell density increases beyond this level, oxygen utilization exceeds oxygen transfer to the liquid phase.

Table 5.1 The influence of temperature on the solubility of oxygen in pure water.

Temperature (°C)	Oxygen solubility in water (mg/l) at 1 atmosphere (101 kPa)
10	10.93
15	9.90
20	8.87
25	8.10
30	7.46
35	6.99
40	6.59

For large-scale culture and/or high cell densities the oxygen demand of the culture can be met only by forced aeration. In practice this is achieved by blowing air through the culture. The efficiency with which oxygen is transferred from air bubbles to the liquid phase principally depends on two functions: the surface area to volume ratio of the air bubbles and the residence time of the bubbles in the liquid. The smaller the bubbles, the greater the surface area to volume ratio and the greater the oxygen transfer. Similarly, the longer the bubbles remain in the liquid the greater will be the amount of oxygen which will diffuse from the bubble into the liquid. One way of decreasing bubble size is to introduce the air through a sparger with multiple small orifices rather than through a single large-bore orifice. A second way is to agitate the culture broth vigorously with a stirrer. There are two additional advantages of agitation. The residence time of the bubbles is

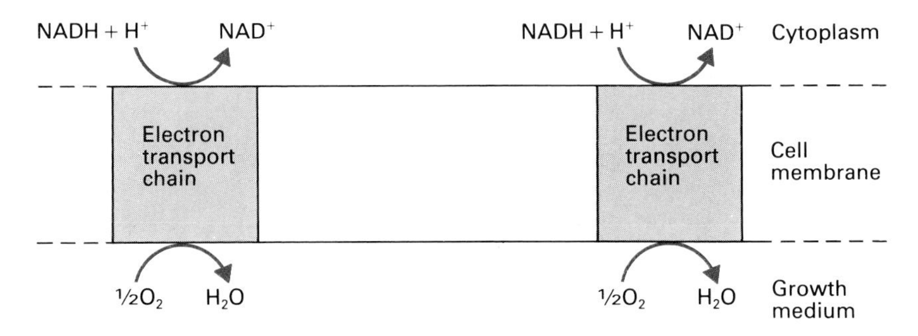

Fig. 5.2 The need for dissolved oxygen. Only oxygen in solution can interact with the cell membrane-bound electron transport chain.

increased because they now follow a tortuous path rather than a straight path to the liquid surface. In addition, agitation facilitates oxygen transfer from the bulk liquid to the cell surface. Agitation is usually effected with a Rushton-type impeller (Fig. 5.3) since this is more efficient at bubble size reduction than a conventional marine propeller. The number of blades on the impeller and their dimensions relative to the fermenter dimensions greatly affect the efficiency of oxygen transfer.

BASIC FERMENTER DESIGN

From the above it should be apparent that the basic fermenter consists of a closed vessel fitted with an air inlet and an agitator. However, many other features are required and some of these are outlined below.

Baffles Addition of baffles to the vessel walls can improve the efficiency of oxygen transfer by increasing the turbulence of the agitated culture medium.

Anti-foam control Agitation and aeration of the culture medium can result in excessive foaming, particularly at high cell densities and/or if complex growth supplements such as yeast or soya extract are present. Foaming can be detected by means of a simple probe placed above the culture medium. When the foam touches the probe it completes an electrical circuit and this activates a pump connected to a supply of an anti-foaming agent.

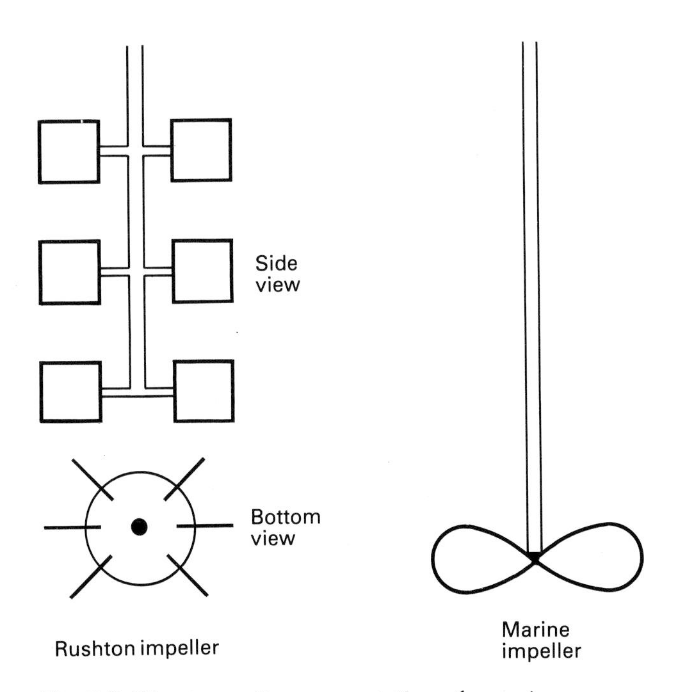

Fig. 5.3 Diagrammatic representation of a marine impeller and the six-bladed Rushton impeller.

pH control The metabolism of most microorganisms results in a change in the pH of the culture medium with a concomitant change in the physiology of the microbe. It is usual practice continuously to monitor the pH of the fermentation broth with a pH probe *in situ* and to maintain a fixed pH by the addition of acid or alkali.

Temperature control Microbes produce heat as they metabolize substrates. Exchange coils inside the vessel and/or a vessel jacket are used to counter this metabolic heat. As the fermentation proceeds the cell density increases, as does the heat output, and cooling water is circulated through the coils or jacket. Cooling water is used in an identical way to reduce the temperature of the fresh medium after it has been sterilized.

Addition ports Provision must be made for the addition of the culture inoculum and media requirements.

CURRENT DESIGN

The above requirements are embodied in the basic stirred tank reactor design of fermenter as shown in Fig. 5.4. The material of fabrication is stainless steel and this must be of a high grade if it is not to corrode or leak toxic metal salts into the growth medium. The design of the fermenter shown in Fig. 5.4 is little changed from that developed in the 1940s for penicillin production, although the engineering is now to a far higher standard. The biggest change has been made to the agitators. Originally the agitator shaft was inserted through the roof of the vessel and the seal between the shaft and the vessel was a simple stuffing box; current practice is to use mechanical seals (see below) and to use bottom-driven agitators (Fig. 5.5). Since bottom-driven agitator shafts are shorter and are fully immersed in liquid there is much less 'whiplash' of the shaft and consequently fewer mechanical problems.

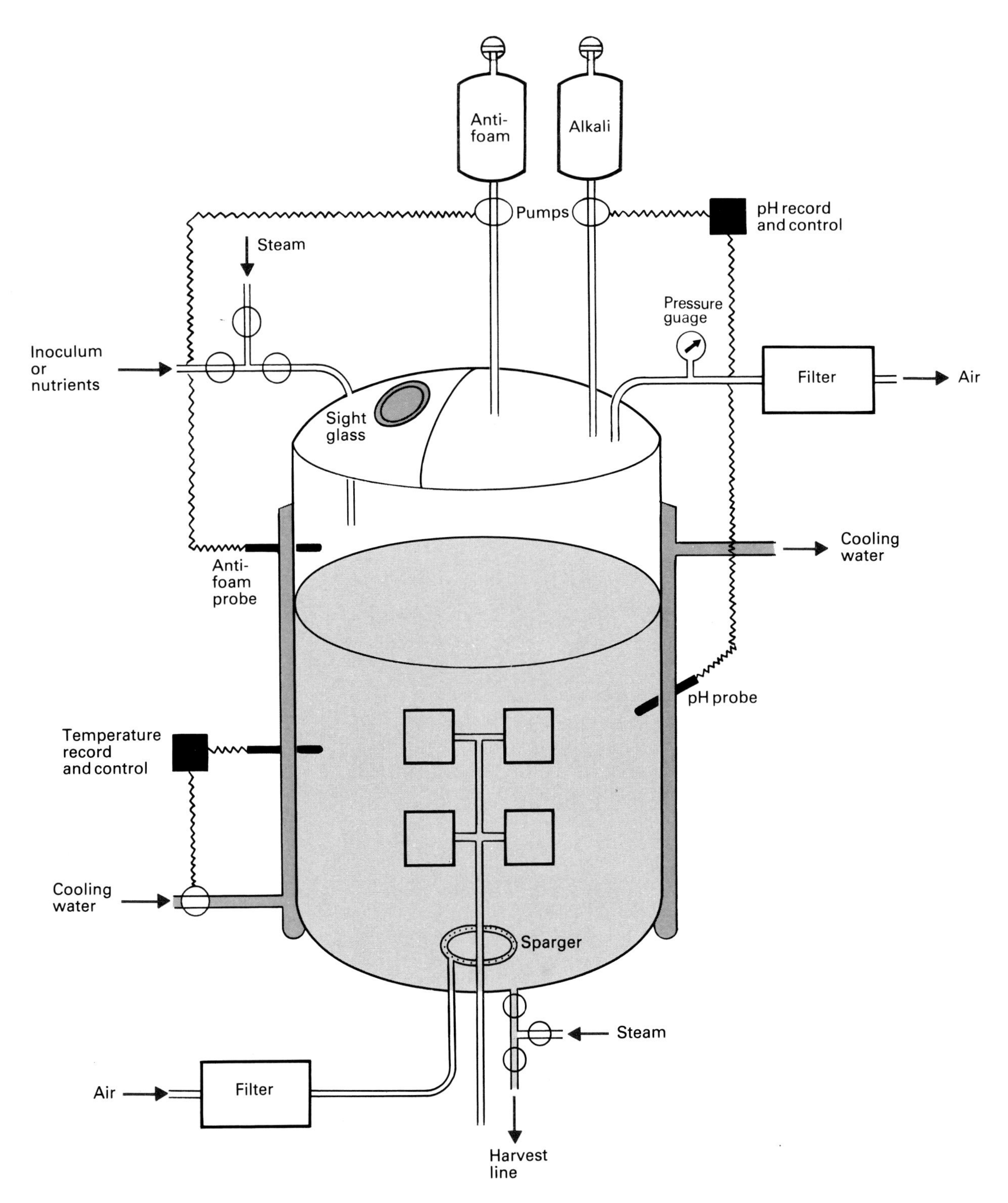

Fig. 5.4 Schematic representation of a stirred tank reactor. For clarity no seal is shown between the agitator shaft and the fermenter body and baffles have been omitted.

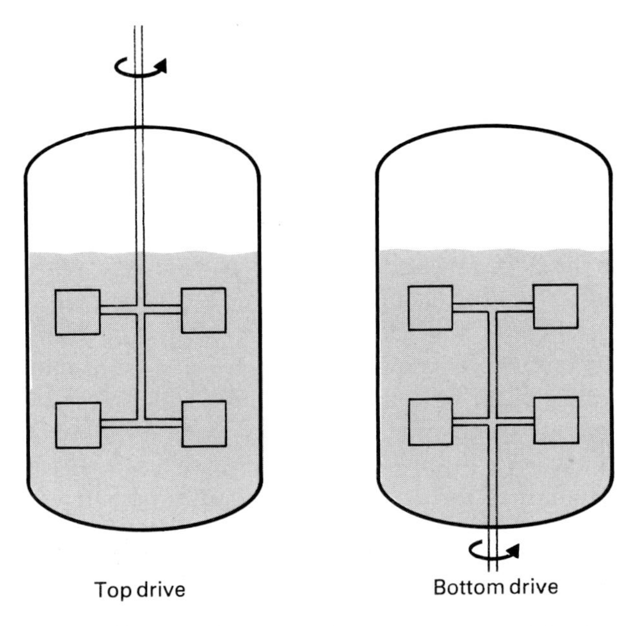

Fig. 5.5 Top- and bottom-driven agitator assemblies for fermenters. Note that the shaft length is shorter with a bottom drive system.

ASEPTIC OPERATION

A key requirement for a successful fermentation is aseptic operation; that is, no contamination of the culture should occur. Thus the entire fermenter and ancillary equipment, as well as the growth medium, must be sterile before inoculation. In addition, the air supplied during the fermentation must be sterile and there must be no mechanical breaks in the fermenter that will allow the ingress of microorganisms.

All the equipment is cleaned with hot water before use and then sterilized *in situ* with steam. For sterilization to be successful the steam must reach all parts of the fermenter assembly, i.e. there must be no pockets of air. This requires judicious equipment design, particularly in the location of valves. In addition, all surfaces must be smooth and thus welds, etc., must be polished; if any rough surfaces are present, however small, they can act as reservoirs of contaminating microorganisms. The growth medium can be heat sterilized in the fermenter itself by passing steam through the cooling coils and jacket. Alternatively, the medium can be prepared in a separate vessel and passed through a continuous sterilizer, in essence a series of stacked hot plates, *en route* to the fermenter. Similarly, additives such as antifoaming agents must be sterilized. The concentrated acids or alkalis used for pH control do not need to be sterilized.

The air supplied to the culture medium is sterilized by filtration. Many different kinds of filter material have been developed but in most instances the microbes in the air are not removed by size exclusion. Rather, the air follows a tortuous path through the filter matrix and the microbes are entrapped. Although bacteria and fungi are removed by filtration, bacteriophages in the air may not be and their presence in the fermenter can result in wholescale loss of the culture. Careful choice of filter design and air flow rate can minimize the problem as can the choice of air compressors. Some compressors generate sufficient heat to inactivate any bacteriophages present. It does help if a fermentation plant designed for large-scale culture of *Escherichia coli*, for instance, is not located downwind from a sewage treatment plant or a farmyard: both are rich sources of coliphages! Air leaving the fermenter may also need to be sterilized; this will certainly be necessary if recombinant organisms are being grown. This is not only a safety requirement but it prevents key microbial strains from being released into the environment and hence becoming freely available to competitors. Again, filtration is used.

The point at which the agitator drive shaft enters the vessel is a potential source of contamination. In the earliest fermenters stuffing boxes (Fig. 5.6) were used to minimize microbial entry. As the name implies, these were boxes stuffed with glass wool and phenolic grease. More recently, double mechanical seals (Fig. 5.6) have been introduced and the moving parts are lubricated with condensed steam to maintain sterility.

OPERATION OF FERMENTERS

The starting point for any fermentation is a clean vessel which is sterilized and charged with sterile medium. The fermenter then has to be inoculated and the size of the inoculum generally is of the order of 1–10% of the total volume of medium; if it is any smaller, there may be a prolonged lag period before growth commences and the fermentation period will be unduly prolonged. An Erlenmeyer flask culture can be used as the inoculum for a small fermenter with a capacity of 10–20 l. Once the fermenter volume increases beyond this it is necessary to prepare the inoculum in a smaller fermenter. For a very large production fermenter a fermenter train

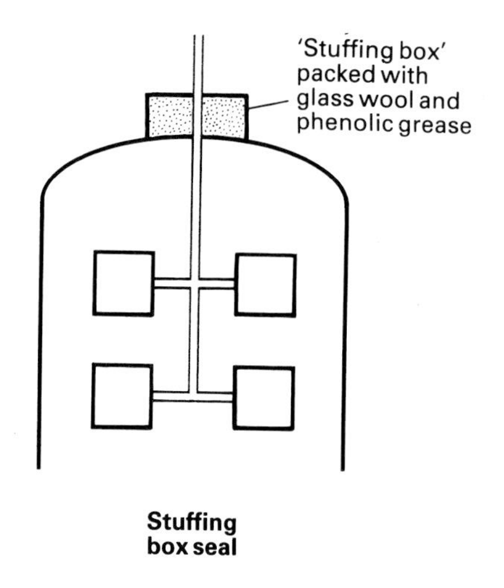

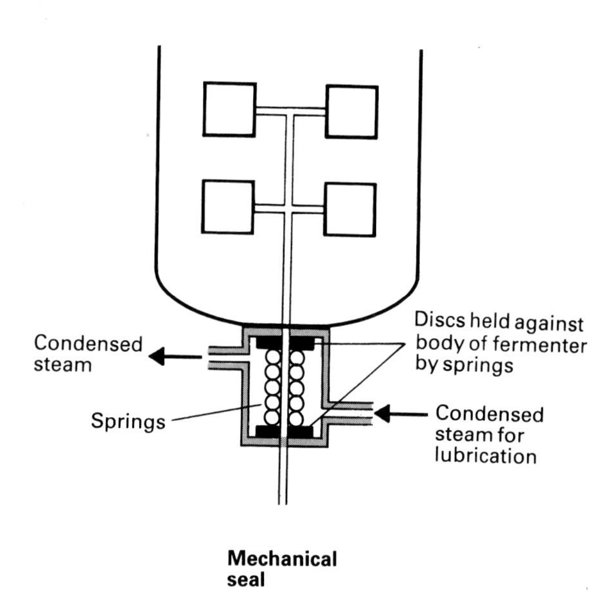

Fig. 5.6 The original 'stuffing box' seal and a modern mechanical seal used to prevent ingress of microorganisms where the agitator shaft passes through the body of the fermenter. Mechanical seals can be used with either top- or bottom-driven agitators. Stuffing boxes cannot be used with a bottom drive system for the vessel contents would leak out.

(Fig. 5.7) may be needed to provide the inoculum. This greatly increases the capital requirements of the process.

There are two major problems associated with the scale-up of fermentations. The first of these is culture stability. As outlined in Chapter 1, the classical method of strain development is repeated mutagenesis and culture selection. This eventually yields an enfeebled strain which overproduces the desired metabolite. When such high-yielding strains are cultivated they can be overgrown by more rapidly growing revertants which give reduced product yields. The chances of this happening increase as the scale of the fermentation increases. Ideally synthesis of the product should be 'switched off' until the final production fermenter is reached. With classical overproducing strains this is seldom possible. There is much less of a problem with genetically engineered strains because enzyme or protein synthesis can be controlled by judicious choice of promoter (see p. 14).

A second problem associated with scale-up is that many of the physical and chemical parameters which influence the behaviour of a microorganism are changed as the scale of operation is changed (Table 5.2; also see Box, p. 56). Thus operating conditions which are optimal for product formation at, for example, 100 l are not necessarily optimal at a much larger scale. Unfortunately it is too expensive and time consuming to do a large number of fermentations at the production scale in order to find the ideal conditions. A number of techniques are used to overcome this problem, albeit not very satisfactorily, but they will not be discussed within the scope of this chapter. Such scale-up problems are not restricted to overproducing-organisms developed by classical mutation and selection procedures. Although many genetically engineered organisms are immune to these problems, again because of the use of controllable promoters, filamentous organisms are not. This is because organisms such as actinomycetes and filamentous fungi are shear sensitive and this, in turn, affects their morphological filament size, the viscosity of the culture medium and the oxygen transfer coefficient.

Table 5.2 Variation in physical parameters on scale-up.

Scale-up criterion (parameter held constant)	Change on 1000-fold scale-up		
	Power	Power/ unit vol.	Impeller speed
Equal power/unit volume	10^3	1	0.22
Equal impeller speed	10^5	10^2	1.0
Equal impeller tip speed	10^2	0.1	0.1

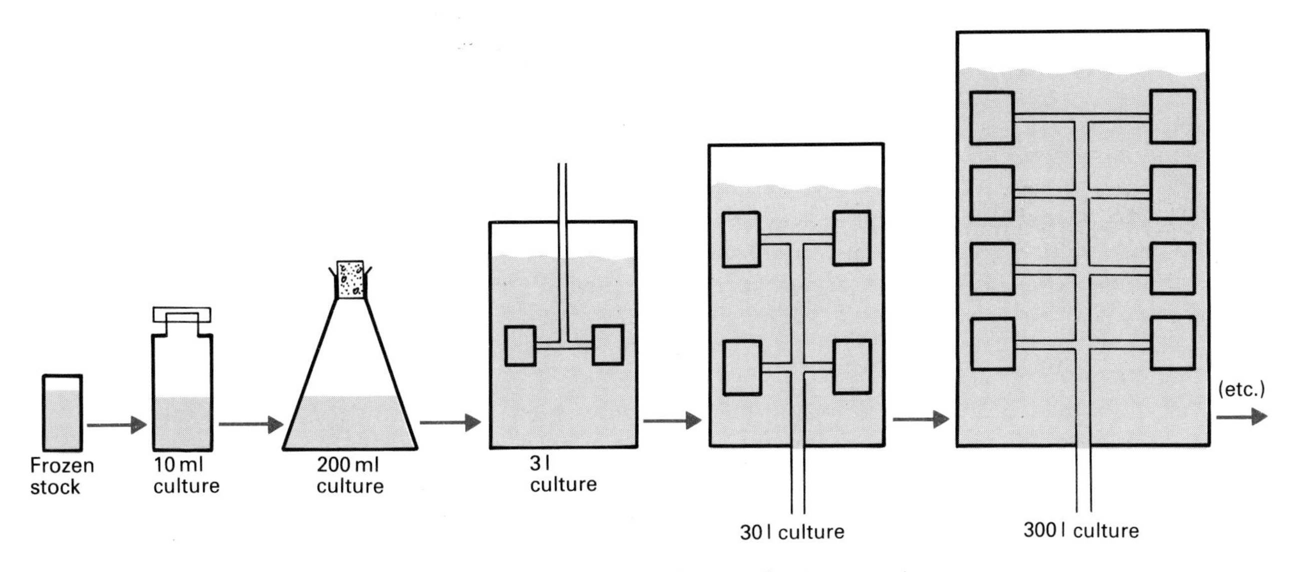

Fig. 5.7 A fermenter 'train' used to provide a large inoculum for a production vessel.

CONTROL SYSTEMS

Many commercial aerobic fermentation processes are difficult to optimize and are not particularly reproducible. One cause of these problems is the use of complex nutrients such as starch, vegetable oil, soya meal, molasses and corn steep liquor. These nutrients are not chemically defined and there is substantial batch to batch variation which can significantly affect product yield. There even can be substantial variation in contaminating substances in the bulk inorganic salts, e.g. phosphate, nitrates, etc., used in media preparation. These problems are compounded, for the reasons given in the previous section, if filamentous organisms are used. Ideally the fermentation technologist would like continuously to monitor the intracellular activities of the cell and to build up a profile of a standard fermentation. Then, if any deviation from the standard profile is detected in subsequent fermentations, corrective action may be taken by altering the environmental conditions.

If such *real time* process control is to be developed, there are two prerequisites. The first is suitable computer systems and the appropriate hardware and software is now available. The second is the ability to measure intracellular activities on-line. To date the only such activity which can be measured on-line is the $NAD^+/NADH$ ratio and this is done fluorimetrically. Although there are assays for the ATP concentration inside the cell, e.g. the luciferase assay, these have to be done off-line. On-line measurements which are readily available are temperature, pH, dissolved oxygen and exhaust gas composition. The pH value and dissolved oxygen content of the medium provide essential information for alarm purposes, i.e. any deviation must be corrected, but would give a misleading picture of the process if used for control purposes. This is particularly true for the dissolved oxygen reading since this is a function of both medium viscosity and chemical composition and these change radically as the fermentation progresses.

Exhaust gas analysis can be done on-line and if a mass spectrometer is used much valuable physiological information can be obtained, e.g. respiratory quotient and the oxygen uptake rate. Thus, if an undesirable change in the oxygen uptake rate is detected, it is possible to decide which of the following strategies would have the least undesirable impact on the fermentation: alteration in the nutrient feed rate, the agitator speed, the air supply rate or the gas pressure inside the vessel.

BATCH VERSUS CONTINUOUS OPERATION

In batch processing the fermenter is filled with medium, sterilized, inoculated and the fermentation allowed to proceed. During the time course of the fermentation nutrients, acid or alkali, antifoam and air are supplied and product gases are removed. On completion, the fermenter is emptied and cleaned and prepared for a fresh run. In some instances

Definition and significance of some terms used by biochemical engineers

Aspect ratio The ratio of height to diameter of a fermenter. For a conventional stirred tank the ratio is between 1:1 and 4:1.

K_La This is the oxygen transfer coefficient. The rate at which oxygen is transferred from a gas bubble to the liquid phase can be expressed as:

$$\text{Rate} = K_L a \ (c^* - c_L)$$

where a is the surface area of the bubble, c^* the concentration of oxygen in the bubble (equivalent to the partial pressure of oxygen), c_L is the dissolved oxygen concentration, and K_L a constant of proportionality. Since it is not possible to measure a with any reliability, it is usual to measure K_La. Note that if the value of c^* is increased by raising the air pressure inside the fermenter, the rate of oxygen transfer to the liquid will increase.

Liquid hold-up The volume increase that occurs when gas is bubbled through a liquid.

Metre-cubed Slang way of referring to cubic metres of fluid. For small volumes it is convenient to measure volume in litres but for larger volumes measurement is in cubic metres. One cubic metre is equivalent to 1000 l. Other useful figures are: 1 imperial gallon = 4.55 l; 1 US gallon = 0.83 imperial gallons = 3.79 l.

Metric ton There are 1000 kg (equivalent to 2205 lbs) in a tonne or metric ton whereas there are 2240 lbs in an imperial or US ton.

Newtonian fluid Shear forces are generated when two contiguous liquid or solid surfaces slide in opposite directions. A Newtonian fluid is one whose rate of shear is proportional to shear stress. The constant of proportionality is the fluid viscosity. Most fermentation media exhibit Newtonian fluid behaviour prior to inoculation and even when there is substantial growth of bacteria and yeast. Culture fluids containing filamentous microorganisms exhibit non-Newtonian fluid behaviour because they are much more viscous than the cell-free culture fluid. This complicates calculations of how much power needs to be imparted to the agitator.

Oxygen uptake rate The rate at which cells utilize dissolved oxygen.

Power per unit volume The power used to agitate a given volume of liquid. If the same power per unit volume is used to agitate two different sizes of vessel, it can be shown that the tip speed (q.v.) of the impeller will be greater in the larger vessel. This, in turn, will increase the shear forces on the microorganisms present.

Tip speed The speed at which the edge of an impeller moves through a liquid (or space). Numerically it can be expressed as:

$$\text{tip velocity} = \text{rpm} \ (2\pi r)$$

where r is the distance of the impeller tip from the centre of rotation. Thus at any set rpm the greater the diameter of the impeller, the faster the tip moves through the liquid and the greater the shear force exerted on the liquid and any microorganisms it contains. Thus damage to mycelia increases as fermenters increase in size even if the rpm of the agitator is held constant.

Volumetric air flow rate The volume of air supplied to a culture in a given time relative to the volume of that culture. Thus at a flow rate of 1 VVM, one volume of air is supplied to an equal volume of culture fluid in one minute. If the volume of liquid in the fermenter were 100 l, then 100 l of air would be supplied in one minute.

Working volume The maximum volume of a fermenter to which culture fluid can be filled before aeration and agitation commence. Space must be left for the volume expansion (liquid hold-up) which will occur on aeration and for any foam which will form.

this turnaround time between fermentations can be prolonged.

In continuous fermentation raw materials are supplied to the vessel at a rate volumetrically equal to that at which product and spent medium are withdrawn (Fig. 5.8; see also Box below). One advantage of continuous culture is that provided long runs, e.g. of many weeks duration, can be achieved without contamination then there is no significant turnaround time. In this way the productivity of a particular vessel is increased. The extent of this increase will depend on the ratio of fermentation time to turnaround time: the smaller the ratio, the more advantageous is continuous operation. A second advantage is that the growth rate of the cells can be maintained at that which is optimal for product formation.

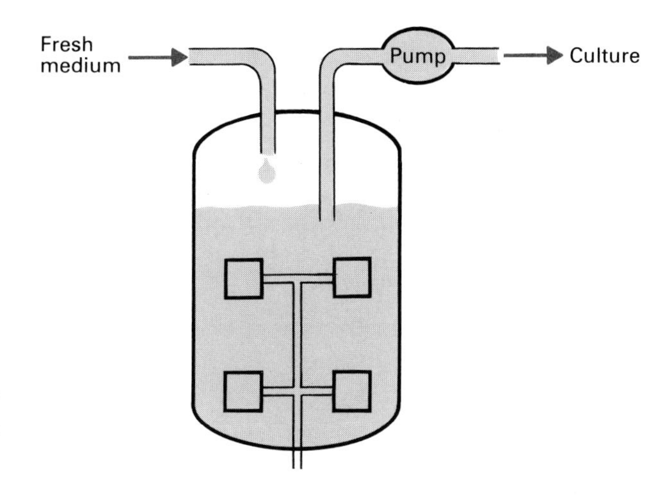

Fig. 5.8 Addition of fresh medium and removal of culture fluid in a continuously operated fermenter in the chemostat mode. The arrangement shown is just one of many used to ensure a uniform volume of culture fluid.

Theory and practice of continuous culture

If unicellular microbes are growing exponentially in a nutrient medium then the rate of change of cell number (n) with time (t) will be proportional to the number of cells. That is:

$$\frac{dn}{dt} = \mu n$$

where μ is the constant of proportionality, otherwise known as the growth rate constant. Eventually some nutrient will become limiting, the growth rate will decrease and finally all growth will cease. In continuous cultures fresh nutrients are constantly added to the medium and an equal volume of culture removed, i.e. the total culture volume remains the same. Nutrient addition can be controlled in two ways. In *turbidostats* the turbidity of the culture is monitored constantly and when it rises above a pre-set value fresh medium is added. In *chemostats* growth is controlled by limiting the supply of a critical growth factor, i.e. some nutrient is limiting. If fresh medium is added to a chemostat at a rate of F l/h and the chemostat has a volume of medium V l, then the rate of loss of organisms from the fermenter is given by:

$$-\frac{dn}{dt} = \frac{F}{V}n = Dn.$$

The ratio F/V is known as the dilution rate (D) and has the dimension of reciprocal hours (h^{-1}). At the steady state, dn/dt should be zero, in which case $\mu = D$, i.e. the rate of growth is identical to the dilution rate, which is a function of the rate at which fresh growth medium is added to the vessel. Clearly, if D is allowed to exceed μ, cells will be lost at a rate faster than they can grow and *washout* will occur. A useful approximation is that when the dilution rate is 0.1 h^{-1} the cells have a doubling time of 7 h and when the dilution rate is 0.7 h^{-1} the cells have a doubling time of 1 h.

In chemostat culture all nutrients except one are present in excess and the level of the limiting nutrient determines the cell density. The nutrients that can be used to limit growth include the sources of carbon, nitrogen, phosphorus, sulphur or metal ions. Different limiting nutrients affect the cell physiology in different ways.

The advantage of turbidostat and chemostat culture over batch culture is that the cells are in a constant environment. However cells in turbidostat culture grow at the maximum rate permissible with the selected medium; cells in chemostat culture are always starving for some nutrient and are not growing at the maximum rate possible except when the value of D approaches that of μ.

Continuous culture is ideally suited for the production of biomass or simple metabolites such as ethanol whose synthesis is proportional to cell density. It is not suitable for the production of other metabolites such as amino acids and antibiotics whose synthesis is not associated with growth. In addition, the strains used in these processes are too unstable for continuous operation (see p. 54). The most widespread use of continuous culture is in the effluent disposal industry (see p. 82) where waste organic materials (substrate) are converted into microbial cells (biomass).

NOVEL REACTOR DESIGNS

The stirred tank reactor (Fig. 5.4) is the workhorse of the fermentation industry because of its reliability and versatility. Nevertheless some novel designs have been introduced in recent years. Considerable power is needed to drive the agitators used on large fermenters and to compress the air supplied. This leads to high utilities costs, principally for electricity. One way of reducing the power requirements is to use the air supplied to agitate the culture. In an *airlift* fermenter (Fig. 5.9) the air is introduced at high velocity at the bottom of the vessel. The use of a draught tube enables the liquid to circulate with considerable turbulence. For efficient oxygen transfer it is necessary for the fermenter to have a height to diameter ratio (10:1–40:1) that is much greater than with a conventional stirred tank reactor (1:1–4:1).

As a gas bubble ascends a column of liquid the hydrostatic pressure on it decreases and the volume of the bubble increases. Since there now is a decrease in the surface area to volume ratio there is a decrease in the rate of oxygen transfer from the gas phase to the bulk liquid. Furthermore, in a well

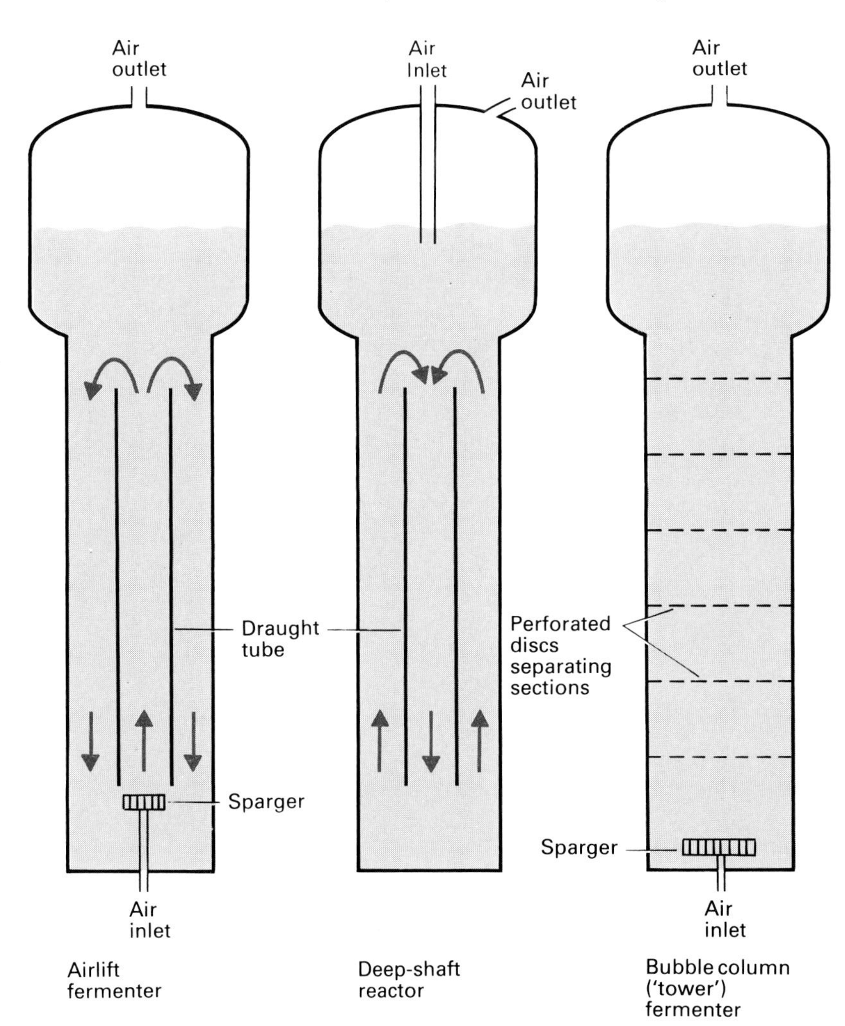

Fig. 5.9 Some novel reactor designs (see text for details). Of the three designs illustrated, only the airlift fermenter is used in both the batch and continuous modes; the other two are always used in the continuous mode.

agitated fermenter over 50% of the oxygen supplied is utilized. Thus, as the bubble ascends the column the partial pressure of oxygen within it will decrease and this too will decrease the rate of oxygen transfer to the liquid. A novel reactor design which overcomes these disadvantages is the *deep-shaft reactor* (Fig. 5.9). Here the air is introduced at high velocity at the top of the reactor and forces liquid downwards. Although oxygen is utilized during the downward passage the bubble size will diminish because of the hydrostatic pressure and should give rise to a more uniform oxygen transfer rate.

Both the airlift fermenter and the deep-shaft reactor can be operated continuously. With both of them the entire vessel contents are essentially homogeneous, both spatially and temporally. By contrast, in a bubble column reactor (Fig. 5.9), which is a variant of the airlift fermenter, the contents of the vessel are not homogeneous. Since there is no mixing, the medium moves through the fermenter as a 'plug' and its composition varies spatially. By dividing the fermenter into sections it is possible to vary the physicochemical environment at will, e.g. pH 7.0 and 30 °C in one section and pH 5.0 and 40 °C in another section.

DOWNSTREAM PROCESSING

Once a fermentation is complete, it is necessary to recover the desired end product. At a minimum this will involve separation of the cells from the fermentation broth but also may include purification of metabolites with or without cell disruption. Such operations are referred to as *downstream processing*. Where the desired product is heat labile it is necessary to cool the fermentation broth prior to cell removal. This can be achieved either by passing ice-cold water through the fermenter coils and/or jacket or by passing the fermenter contents through a heat exchanger.

Some cells rapidly settle out of suspension once aeration and agitation of the fermenter broth cease and this settling may be assisted by the addition of flocculating agents. Many of these flocculating agents are polyelectrolytes which neutralize the charge on the microbial cell surface which keeps cells in suspension by electrostatic repulsion. Where cell settling occurs, or can be induced to occur, then cell removal is relatively inexpensive for the fermentation broth can be decanted or filtered off. Flocculation is an essential property of yeasts used in the production of alcoholic beverages. Usually yeast sedimentation occurs towards the end of the fermentation but in traditional ale production in the UK the yeast cells float to the surface and are skimmed off.

Where cell settling does not occur, cell removal can be effected by centrifugation. Efficient cell removal is favoured by large particle sizes, a large density difference between the particle and the fluid, a low fluid viscosity, a large centrifuge radius and a high angular velocity. The particle size and density and the viscosity of the fermentation liquor usually are outside the control of the fermentation technologist. Additionally, the radius of the centrifuge bowl cannot be increased indefinitely since the mechanical stress on the bowl increases with the square of the radius and safe limits are quickly reached. Thus for large volumes of fluid centrifuges which can operate in a continuous-flow mode are required. Many different designs are available but only those that can operate at high angular velocities (Fig. 5.10) are of use for removing cells or cell debris. When operating continuously these centrifuges are more efficient at particle removal when the liquid to be clarified is flowing through them slowly. This is because the time that each particle is exposed to the centrifugal force is increased.

An alternative to centrifugation is *ultrafiltration*. The term ultrafiltration describes processes in which particles significantly greater in size than the solvent are retained when the solution is forced through a membrane of very fine pore size, usually less than 0.5 μm. Not surprisingly with such fine pore sizes there is tremendous resistance to the passage of liquid and acceptable flow rates are achieved only by the application of pressure and by the use of filter designs that have a high surface area to volume ratio. Extracellular proteins may or may not be retained by the membrane depending on the membrane pore size.

The clarified fermentation liquor will contain microbial metabolites and extracellular enzymes and many methods are available for their recovery including ion exchange chromatography, precipitation and solvent extraction. The use of ion exchange resins is particularly desirable since they can be used for concentration as well as having high capacity and being cheap and reusable.

If the desired product is retained within the cell,

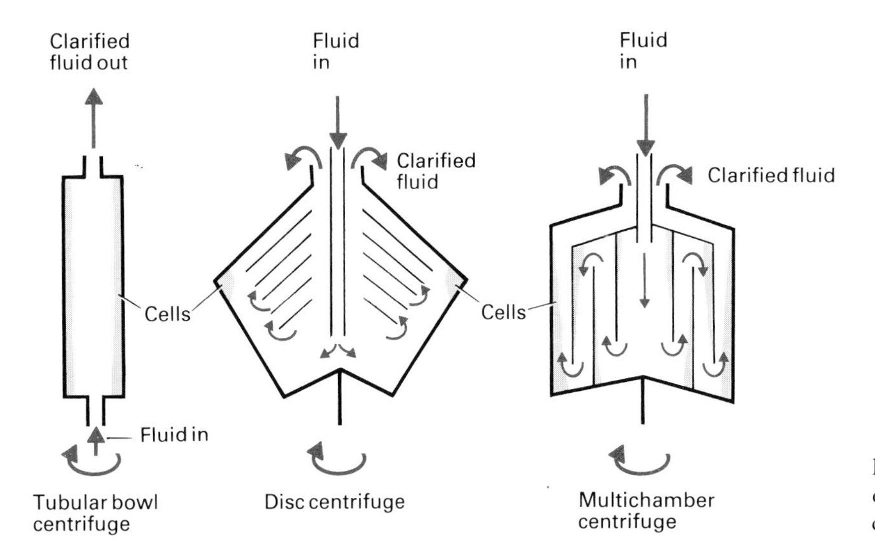

Fig. 5.10 Three commonly used designs of continuous-flow centrifuge.

then cell disruption will be necessary to effect its release. Although many methods of cell disruption are used in the laboratory most of them cannot be used on a large scale. Two which are suitable for large-scale use are high pressure homogenization and the use of high speed bead mills. In high pressure homogenization cell suspensions move from a high pressure environment to a low pressure environment by passage through a fine orifice. The cells rupture by explosive decompression and the process is facilitated by their impaction at high velocity on components of the homogenizer (Fig. 5.11(a)). Cell disruption, as measured by release of protein from the cells, is increased with increasing temperature and pressure and is independent of cell concentration. The desirability of high pressure units is best illustrated by the fact that the efficiency of disruption is approximately proportional to the cube of the pressure.

High speed bead mills are agitators consisting of a central shaft on which a set of circular discs is eccentrically mounted to form a helical array (Fig. 5.11(b)). The tank is filled with very small (0.45 mm) diam.) glass beads prior to charging with the cell slurry. It is the abrasion caused by the rapid motion of these beads which ruptures the cells. The efficiency of disruption increases with increasing agitator speed and bead loading and decreasing bead size, temperature and cell concentration.

After cell disruption the cell debris needs to be removed. As with removal of intact cells the most widely used methods are centrifugation and ultrafiltration.

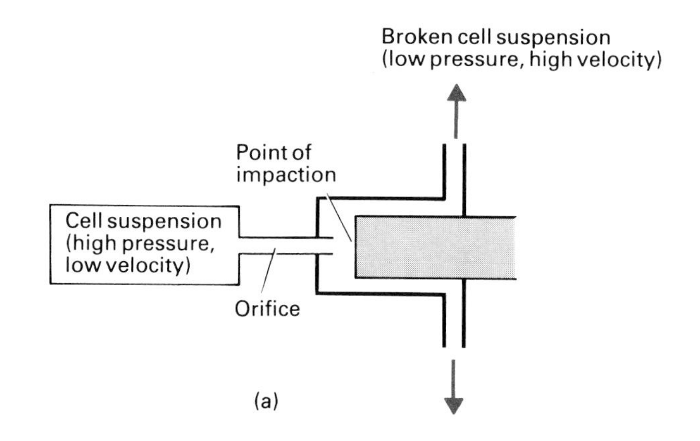

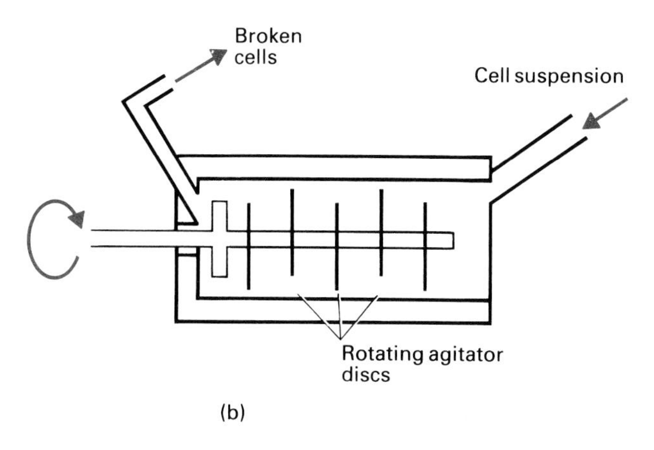

Fig. 5.11 Schematic representations of equipment used for continuous disruption of cells (see text for details). (a) High-pressure homogenizer. (b) High-speed bead mill (beads not shown for clarity).

SAFETY CONSIDERATIONS

There have been few safety concerns associated with the more traditional large-scale commercial fermentations. This is because most of the organisms used are non-pathogenic. In recent years, however, there has been considerable debate about the potential hazards associated with genetically engineered organisms. These hazards are purely conjectural and the evidence to date suggests that recombinant organisms are no more dangerous than their traditional counterparts. Nevertheless, it is prudent for the fermentation technologist to take heed of public concern when designing production facilities. In actual fact there is little hazard associated with the fermentation itself for a vessel designed to keep contaminating organisms out also serves effectively to retain potential pathogens! Design features such as double mechanical seals on the agitator shaft and filters on the exhaust gas outlet are particularly useful. There are a number of specific modifications which can be made to decrease any risk; for example, careful design of sampling ports can eliminate aerosol formation and provision can be made to contain any spills which occur. Also, all effluent from the fermenter, including that produced during cleaning, can be passed through a continuous sterilizer or to a kill tank.

Potentially there are more hazards associated with downstream processing than with the fermentation. Figure 5.12 shows some typical unit operations associated with a fermentation. Heat exchangers pose few problems provided care is taken to prevent freezing of the culture fluid from occurring; this could result in fracture of the exchanger following thermal expansion. The centrifuges used are likely to be of the continuous-flow type and these are notorious for generating aerosols in the immediate environment. Hermetically sealed models are available and should be used and all vents on them should be fitted with microbiological filters. Finally, all the equipment should be connected with high quality piping, including high quality welding, and it should be possible to sterilize all of it.

Further reading

GENERAL

Pirt S.J. (1975) *Principles of Microbe and Cell Cultivation.* Blackwell Scientific Publications, Oxford.

Wang D., Cooney C.L., Demain A.L., Dunnill P., Humphrey A.E. & Lilley M.D. (1979) *Fermentation and Enzyme Technology.* John Wiley & Sons, New York.

SPECIFIC

Buckland B.C. (1984) The translation of scale in fermenta-

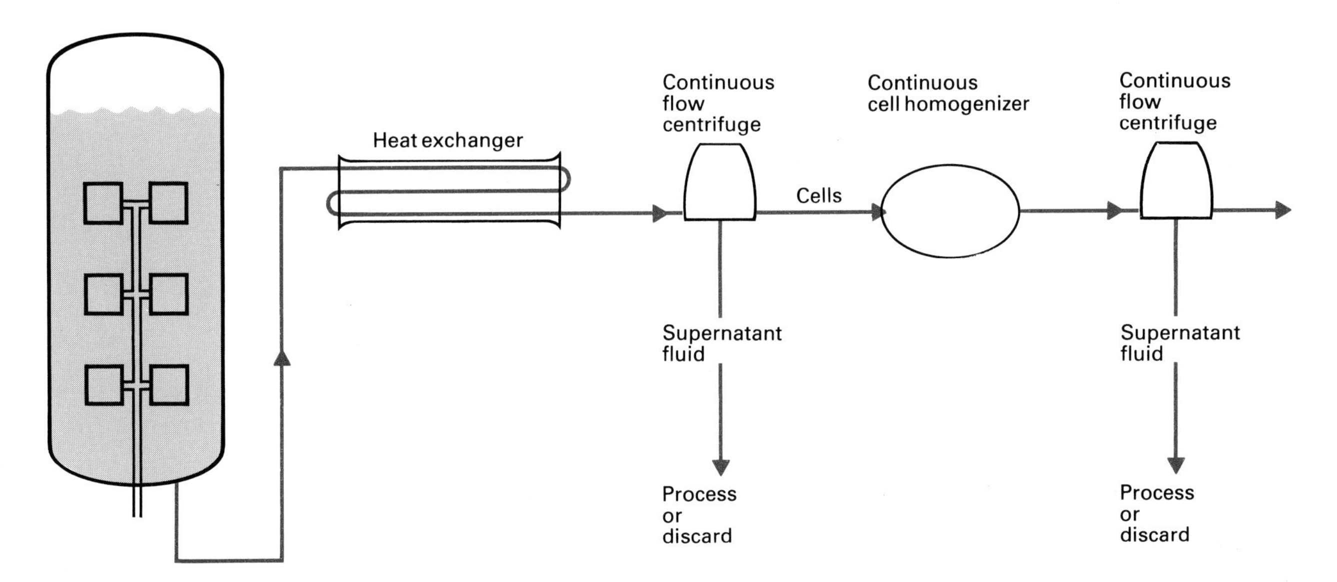

Fig. 5.12 Schematic representation of typical operations in the processing of a fermentation broth. If the desired product is in the broth, the cells are discarded. If the product is inside the cells, the cells may need to be homogenized prior to product extraction.

tion processes: the impact of computer process control. *Bio/technology* **2**, 875–83.

Buckland B.C., Brix T., Fastert H., Gbewonyo K., Hunt G. & Jain D. (1985) Fermentation exhaust gas analysis using mass spectrometry. *Bio/technology* **3**, 982–8.

Tutunjian R.S. (1985) Scale-up considerations for membrane processes. *Bio/technology* **3**, 615–26.

Van Brunt J. (1986) Fermentation economics. *Bio/technology* **4**, 395–401.

6/The Commercial Exploitation of Microorganisms

INTRODUCTION

Microorganisms have been subjugated by man in many different ways. These encompass the production of industrially important materials including fine chemicals (e.g. pharmaceuticals) and bulk chemicals, the manufacture of single-cell protein from diverse substrates, the processing or recycling of waste materials, and coping with what is loosely called the energy crisis. One convenient way of subdividing the different microbial processes is shown in Table 6.1 and each of the different categories will be discussed in turn. Many of these processes are not new. On the contrary, they have been established for two, three or even more decades. Currently many of them are undergoing rapid development and this is a direct result of the introduction of recombinant DNA technology. Other processes are being introduced now for the first time for their existence depends solely on the ability to manipulate genes *in vitro*. One could say that the industry is in a 'ferment'!

Production of whole cells (biomass)

There are a number of instances in which microbial cells themselves are the desired end-product and these are listed in Table 6.2. Most of them have rather specialized applications, e.g. the production of *Penicillium roquefortii* spores. The characteristic flavour of blue cheeses is due to the metabolic products of *P. roquefortii* growing in the milk fats. Strain selection has resulted in the identification of several isolates with particularly desirable properties for cheese manufacture. These include the ability to produce adequate amounts of low molecular weight ketones and alcohols, which are responsible for the characteristic flavour, together with good proteolytic activity which is essential for texture. Production of the spores is limited to a few, small companies but has potential for expansion because cultures grown on milk fats can be used as sources of blue cheese flavouring; naturally produced flavourings are superior to mixtures of purified chemicals because of the flavour subtleties of trace amounts of other metabolites.

Table 6.1 The different biotechnological applications of microorganisms.

1 Production of whole cells (biomass)
2 Production of low molecular weight compounds
   (a) production of primary metabolites
   (b) production of secondary metabolites
   (c) transformation of organic compounds by non-growing cells
3 Production of high molecular weight compounds
   (a) polysaccharides
   (b) lipids
   (c) proteins
4 Processes dependent on general microbial metabolism
   (a) degradation/oxidation of effluents and noxious wastes
   (b) mineral extraction

INOCULANTS

The *Rhizobium* inoculants are of far greater economic value than the blue cheese inocula described above but their use is largely confined to Australia and New Zealand. The reason for this is twofold. First, forage legumes such as clover and lucerne (alfalfa) were not indigenous to these countries but were introduced by the early settlers. Since they were not indigenous crops there was no reservoir in the soil of the *Rhizobium* sp. whose infection of root hairs leads to the beneficial nitrogen-fixing symbiosis which makes legumes so agriculturally important. One way of overcoming this is to add the appropriate rhizobia to the soil along with the seeds. This is in fact what is done. However, in climates such as that of Australia, the added rhizobia tend not to survive from year to year and hence inoculation is required in each planting season.

Despite the economic value of inoculation there probably are fewer than ten companies worldwide,

Table 6.2 Some applications of microbial cells.

Organism	Application
Bacillus thuringiensis and related organisms	Microbial insecticide
Lactobacillus sp., *Streptococcus cremoris* and related species	Starter cultures for the manufacture of dairy products, e.g. yoghurt, cheese
Penicillium roquefortii and related species	Inocula for the production of blue-veined cheeses
Rhizobium sp.	Inoculants for adding to legume seeds to promote nodulation and nitrogen fixation
Pseudomonas syringae	Creation of artificial snow. Ice-nucleation-defective mutants for the prevention of frost damage to crops
Many different organisms	Single-cell protein production

all of them small operations, producing these *Rhizobium* inoculants. Although considerable progress has been made in recent years in understanding the molecular basis for the legume–microbe symbiosis, it is unlikely that gene manipulation will have a significant impact on the development of improved strains for inoculation. Better strains will come about by selection of improved isolates already existing in nature. Although there has been much talk about the development of rhizobia that infect other plants, such as cereals, it is more likely that in the long term the biochemical apparatus for nitrogen fixation will be introduced into plants by manipulation of the plant genome (see Chapter 12).

An organism whose commercial exploitation has only begun recently is *Pseudomonas syringae*. A surface component of this bacterium catalyses the formation of ice nuclei and is finding use in the formation of 'artificial' snow on ski slopes. The presence of this organism on the surface of leaves can promote frost damage to crops, so considerable benefit might be obtained if plants were sprayed with ice-nucleation-defective mutants. Currently this subject is surrounded by controversy regarding the environmental implications and is discussed in more detail in Chapter 14.

INSECTICIDES

The widespread use of agrochemicals has allowed man to achieve unprecedented control over pests and has contributed towards today's high agricultural productivity. This is reflected in the fact that Europe for the first time is a food exporter and has built up extensive reserves of grain and other produce. However the potential toxic effects on man and the environment and the rapid development of resistance in the target pests has led to a reconsideration of the real value of chemical insecticides. Consequently interest is turning to certain microorganisms which are natural pesticides (Fig. 6.1). Although over 100 bacteria, fungi and viruses that infect insects have been described, only a very few are in commercial production (Table 6.3). The most widely used is *Bacillus thuringiensis* which produces intracellularly a proteinaceous crystal toxic to insects. After ingestion by susceptible insects the crystalline toxin dissolves in the alkaline environment of the midgut where it inhibits ion transport leading to cessation of feeding and death. Much still remains to be learned about the mechanism of toxicity; for example the nature and potency of the toxin crystals varies amongst bacterial serotypes and different strains are required to control different pests. Once more is known about the biological basis of toxicity it may be possible to use protein engineering to develop a single strain which is pathogenic to a wide variety of insects. Recombinant DNA technology has already been used to transfer the genetic information for toxin production to a *Pseudomonas* sp. which is commonly found associated with the roots of maize plants. The objective here is to control root-boring pests in the US corn belt. Whether the US Environmental Protection Agency will approve its use remains to be seen (see Chapter 14).

STARTER CULTURES

A major application of microbial cells is the addition of starter cultures to milk for the production of fermented products such as cheese, yoghurt, sour cream, etc. The basic reason for converting milk into products such as cheese is to preserve the valuable nutrients in milk. Thus cheese is made because this affords a means of converting an easily perishable food (milk) into a stabler product (cheese). This is brought about in a variety of ways: a decrease in water content and water activity, a decrease in pH caused principally by the formation of lactic acid, the depletion of food components which

Fig. 6.1 Death of an aphid due to infection with the fungus *Verticillium lecanni*. (Courtesy of Microbial Resources Ltd.)

Table 6.3 Examples of some microbial insecticides in commercial production.

Organism	Used to control
Bacillus thuringiensis	
Kurstak strain	Caterpillars
H-14 strain	Whitefly and mosquitoes
Bacillus popillae	Japanese beetles
Bacillus penetrans	Nematodes (eelworms)
Hirsutella thompsonii	Citrus mite in Florida
Verticillium lecanni	Glasshouse aphids in UK
Metarhizium anisopliae	Spittle-bug in Brazilian sugar cane and pastures
Beauvaria bassiana	Colorado beetle in USSR

could be used by spoilage and pathogenic microorganisms and the production of 'natural' antimicrobial substances such as D-leucine. Another reason for making fermented milk products is to provide both textural and flavour variety in foods.

The use of lactic starter cultures for the manufacture of dairy products is a relatively new business. For thousands of years man depended upon the natural contaminating flora in milk for the synthesis of lactic acid and flavours and the formation of the desired product. However, the widespread thermal processing of milk, e.g. pasteurization, to destroy pathogens such as the causative agents of tuberculosis and brucellosis alters the microbial flora of milk and it no longer naturally undergoes a lactic fermentation. Consequently starter cultures have to be added and these are generally frozen or lyophilized cell concentrates of either pure cultures of lactic acid bacteria or selected mixtures. When these bacteria are added to milk they cleave the lactose ('milk sugar') to glucose and galactose and these are metabolized to lactic acid (Fig. 6.2). The acid coagulates the milk proteins, principally casein, to form a continuous solid curd in which fat globules, water and water-soluble materials are entrapped. The curd is then pressed to expel the fluid whey, shaped and left to mature. During maturation subtle flavours are formed by enzymic conversion of the curd contents and by microbial metabolism.

Growth of lactic acid bacteria in milk is dependent not only on phosphogalactosidase activity but also on protease activity. These bacteria are nutritionally fastidious and require an exogenous supply of amino acids and peptides. In order to grow in milk they must degrade casein and the extent of this proteolysis contributes to the flavour of the final product.

Gene manipulation has much to offer the dairy microbiologist. For example, a major problem in the production of fermented milk products is slow acid formation, which can result in off-flavours, extended process times or complete loss of production. Slow acid formation can be caused by inhibition of the bacteria by antibiotic residues in the milk, loss of the ability to grow on lactose or to degrade casein, or infection with bacteriophages. Now that the genetics of the lactic acid bacteria are receiving much attention the construction of antibiotic-resistant derivatives should not be too difficult. Both the ability to grow on lactose and to degrade casein are plasmid-mediated traits and increasing plasmid stability is feasible. Alternatively, it should

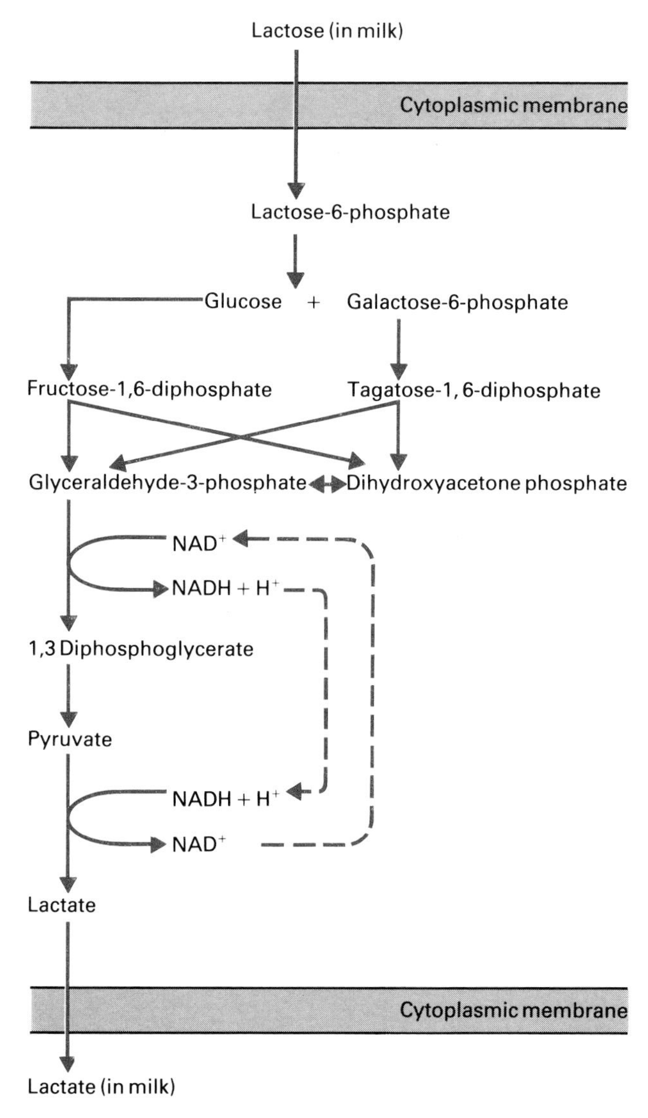

Fig. 6.2 The metabolic interconversions of lactose which lead to lactic acid formation. Not all the intermediates are shown.

be possible to clone the relevant genes and force their integration into the chromosome. Finally, if the number of copies of the plasmid were increased, this could result in more rapid acid formation and hence decreased process times.

Attack by bacteriophages is probably the biggest source of problems for the dairy industry. Bacteriophages abound in milk and the dairy environment generally and their complete elimination is not possible. However, the restriction enzymes found in many bacteria, and which are used for gene splicing, are effective at reducing bacteriophage replication in cells. Indeed, this is how they were discovered. A potential solution to the problem of bacteriophage attack of lactic acid bacteria used as starters is to clone in them a number of different restriction systems. The greater the number of restriction and modification systems inside the cell, the greater the chance that the incoming bacteriophage nucleic acid will be degraded. However the success of such approaches will depend on the attitude of regulatory agencies to the use of recombinant organisms in the preparation of food!

Just as starter cultures are required for the production of fermented milk products, so too are they required for the preparation of fermented meat products, such as salami. The initial bacterial flora of meat supplied for sausage production is more variable today because of modern methods of transporting, pretreatment and slaughter of animals. The starter cultures used are mixtures of lactobacilli and streptococci, to produce lactic acid which acts as a preservative, and micrococci. The nitrate- and nitrite-reducing activities of the micrococci convert myoglobin to nitrosomyoglobin; the latter is essential for red colour formation.

SINGLE-CELL PROTEIN

The term 'single-cell protein' (SCP) refers to cells, or protein extracts, of microorganisms grown in large quantities for use as human (*food*) or animal (*feed*) protein supplements. Although SCP has a high protein content, it also contains fats, carbohydrates, nucleic acids, vitamins and minerals. Its use in food presents a problem because humans have a limited capacity to digest nucleic acids and additional processing is necessary. The animal feed market is more attractive because not only is less processing of the product required but the regulatory process is less stringent.

Over the years a large variety of raw materials has been used for SCP production including carbon dioxide, methane, methanol, ethanol, sugars, petroleum hydrocarbons and industrial and agricultural wastes. These feedstocks have been used with a variety of different microorganisms including algae, actinomycetes and other bacteria, yeasts, filamentous fungi and higher fungi. Commercial success depends, among other things, on the ready availability of cheap substrate and its efficient conversion into cellular protein. It is worth noting that one of the economic realities of biotechnology is that waste products, which by definition are readily available, are cheap only so long as nobody wants them.

For economic, political and technical reasons SCP has not become the important source of food protein that was forecast originally. The developed countries do not need SCP for they have a plentiful supply of high quality protein as a result of the agricultural improvements of the last forty years. Evidence of this is provided by the butter and grain 'mountains' which exist in the EEC and USA. SCP would be a useful food additive in tropical countries where traditional food products are high in carbohydrate and low in protein. This chronic shortage of protein leads to physical and mental deterioration. In countries of the Middle East, and in Africa south of the Sahara, the land is too arid to produce sufficient food of any type regardless of its protein content. Thus in both the arid and tropical regions the production of SCP would be advantageous but the countries concerned cannot afford the initial capital investment. They also lack the technical expertise and support facilities to maintain production once established. One country which might favour SCP production is the USSR since it has the unique combination of technical expertise coupled with poor agricultural productivity and a lack of hard currency.

Even as a feed supplement SCP has not been a success because of cost considerations. Feedstocks, particularly those associated with oil production, generally have increased in price, whereas those from competitive sources of protein, such as soybean meal or fishmeal, have decreased in price. Even if oil prices drop, which is the case at the time of writing, they can just as easily rise again because of world politics. Fluctuating commodity prices do not encourage large capital investment.

SCP production facilities have been constructed in most of the developed countries but many of these are no longer operating for the reasons stated above. The most impressive is the one built by Imperial Chemical Industries in the north of England. In this plant, which has the world's largest continuous operating fermenter, the bacterium *Methylophilus methylotrophus* is grown on methanol and ammonia. The product, which is called Pruteen, contains 80% crude protein and has a high vitamin content. Although twice as nutritious as soybean meal it is not economic to produce.

Recombinant DNA technology has been used in an interesting way to try and improve the efficiency of SCP production. Ammonia must be assimilated for cellular growth and there are two different routes whereby this occurs (Fig. 6.3). *M. methylotrophus* does not possess the enzyme glutamate dehydrogenase and uses instead the ATP-dependent GS/GOGAT pathway. This represents a potential source of methanol wastage. When the glutamate dehydrogenase gene from *E. coli* was cloned and expressed in *M. methylotrophus* there was a 4–7% increase in the efficiency of carbon conversion.

BACTERIAL VACCINES

A clinical use of bacterial cells is for the production of vaccines. Some vaccines, e.g. those for diphtheria and tetanus, are purified exotoxins which have been rendered non-toxic by treatment with heat or formalin. However most bacterial vaccines, such as

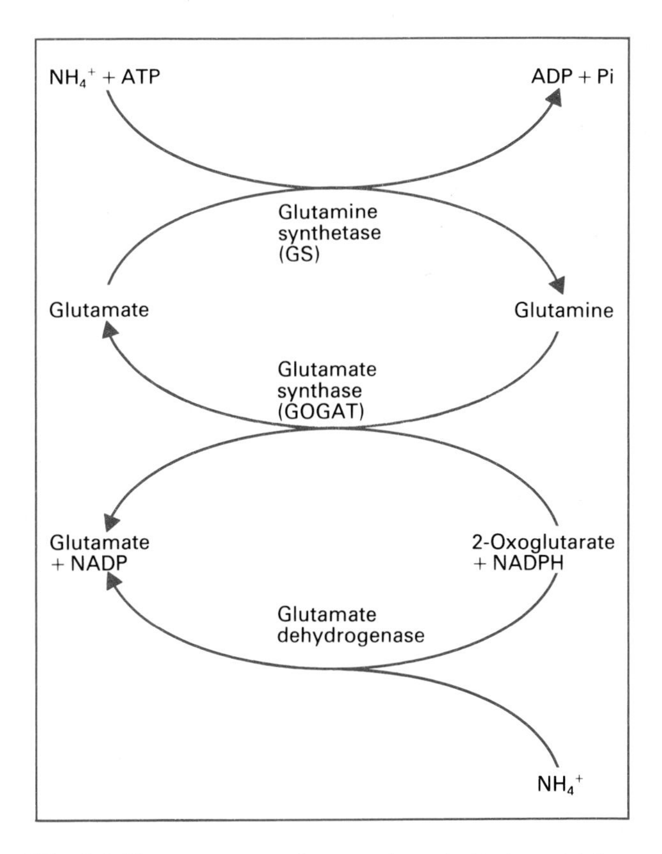

Fig. 6.3 The two routes of ammonia assimilation which are found in microorganisms.

those for cholera, whooping cough, plague, typhoid, paratyphoid fever and typhus fever, are simply suspensions of killed bacteria. After growth the bacteria are killed with heat or chemical agents such as phenol, acetone or formalin. Only the vaccine for immunization against tuberculosis uses live cells, in this case an attenuated strain of *Mycobacterium bovis*. Whereas vaccines prepared from live, attenuated organisms often give lifelong protection, vaccines prepared from killed cells have to be injected repeatedly according to carefully balanced immunization schedules. Recently a novel way of attenuating *Vibrio cholera* has been developed. Using recombinant DNA technology a portion of the gene for cholera enterotoxin was deleted. The modified protein produced is immunogenic but non-toxic. The other development in the production of bacterial vaccines is to use subcellular components of the cell as immunogens; for example, pili on the surface of Gram-negative bacteria are involved in adsorption of the bacteria to animal cells. Purified pili are being tested as vaccines for *Neisseria gonorrhoeae*, *V. cholerae* and enteropathogenic strains of *E. coli* (the cause of 'traveller's tummy').

Production of low molecular weight compounds

Microorganisms are used to produce a wide variety of low molecular weight compounds and some examples are given in Table 6.4. These compounds can be subdivided (Fig. 6.4) into those whose production is associated with growth (*primary metabolites*) and those whose synthesis occurs after growth ceases (*secondary metabolites*). Primary metabolites can be further subdivided into those such as vitamins and amino acids, which normally are produced in quantities only sufficient for cell growth, and those such as ethanol and lactic acid, which are produced in large quantities because they are normal metabolic end-products. Microbes are also used to effect chemical transformations; that is, the desired product is not a normal metabolite of the cell

Table 6.4 Examples of the diversity of commercially useful low molecular weight substances produced by microorganisms.

Compound	Use	Producing organism
Antibiotics		
Colistin	Antibacterial agent	*Bacillus colistinus*
Griseofulvin	Antifungal agent	*Penicillium griseofulvum*
Avermectin	Antihelminthic agent	*Streptomyces avermitilis*
Pharmacologically active agents		
Ergot alkaloids	Vasoconstriction	*Claviceps purpurea*
Cyclosporin	Immunosuppression	*Tolypocladium inflatum*
Mevinolin	Cholesterol reduction	*Aspergillus terreus*
Vitamins		
Cyanocobalamin (B_{12})	Pernicious anaemia	*Propionibacterium shermanii*
Amino acids		
Glutamic acid	Flavour enhancer	*Corynebacterium glutamicum*
Steroids		
11 α-hydroxylation	Production of therapeutically useful steroids	*Rhizopus nigricans*
Other		
Sorbose	Vitamin C production	*Gluconobacter suboxydans*
Indigo	Textile dye	*Escherichia coli*

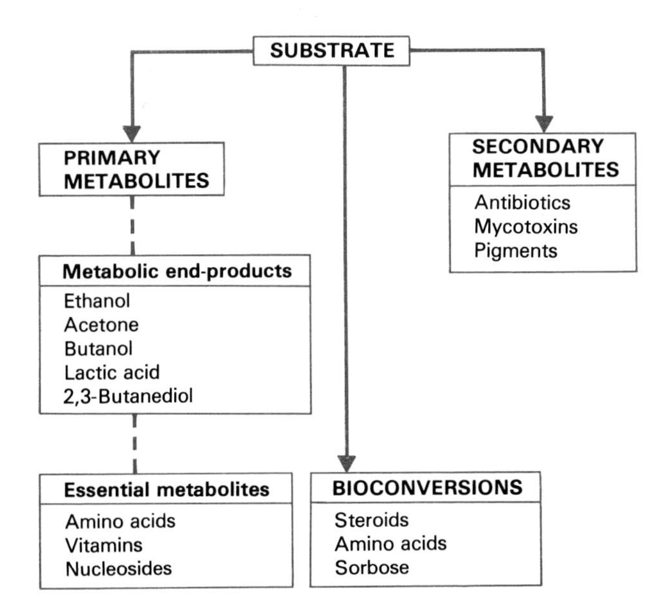

Fig. 6.4 The different classes of low molecular weight compounds synthesized by microorganisms.

but is produced as a result of enzymatic conversion of an unusual substrate added to the culture medium. Often such substrates cannot support growth, they simply undergo *bioconversion*.

METABOLIC END-PRODUCTS

Metabolic end-products such as ethanol, acetone, butanol and lactic acid are the traditional products of the fermentation industry. They are also fermentations in the strictest sense because they are formed as a result of the anaerobic metabolism of sugar substrates. In the absence of oxygen as terminal electron acceptor different cells regenerate NAD^+ from NADH by different means (Fig. 6.5) some of which lead to the formation of commercially important chemicals. Ethanol is different in that it has two functions. Like the others it is a commodity chemical but, more importantly from an economic viewpoint, its production is also the basis of an enormously varied alcoholic beverage industry (Table 6.5). Over 16 billion imperial gallons (20 billion US gallons) of beer alone are produced annually making it the single most important fermentation product.

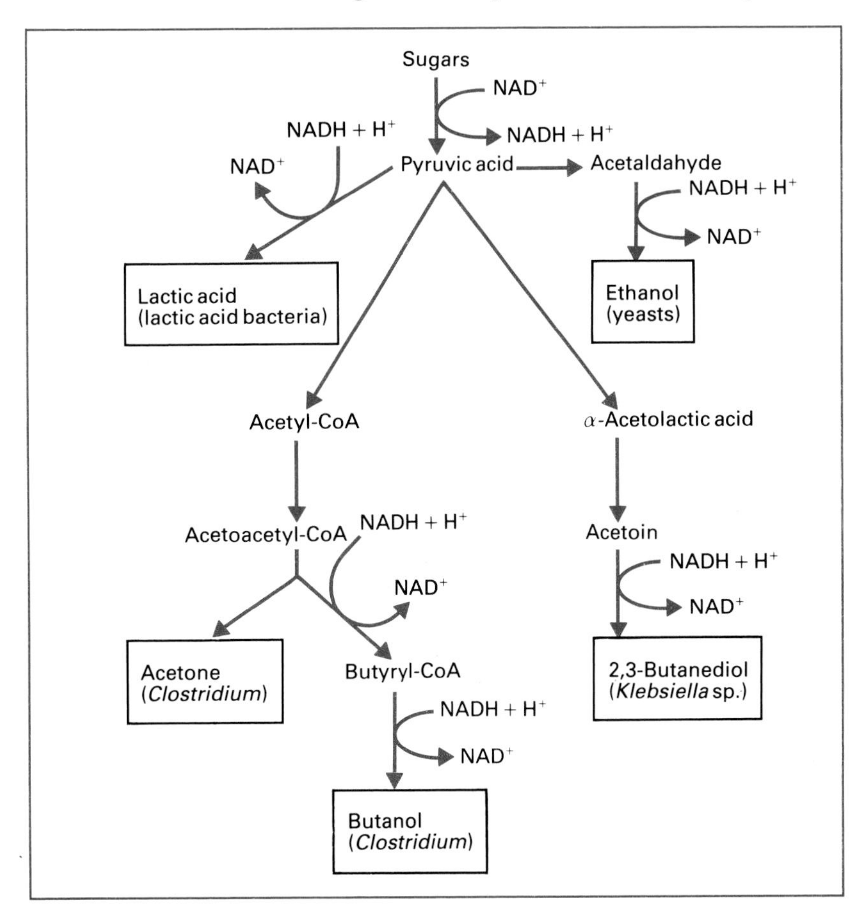

Fig. 6.5 The formation of commercially useful metabolic end-products. Note that pyridine nucleotide co-factors are reduced during the conversion of sugars to pyruvate and subsequently oxidized by further metabolism of pyruvate.

As with SCP, the fermentative production of these metabolic end-products is subject to economic and political pressures. Thus their production declined after the Second World War with the growth of the petrochemical industry which was based on cheap oil supplies. As the price of oil has risen there has been a move back to fermentation as a means of production. Thus the percentage of industrial alcohol production which is based on fermentation has risen in the last ten years from less than 10% of total world demand to more than 20%. This increase in fermentative production is not without its consequences. Much of it takes place in Brazil and is dependent on sugar cane for the provision of substrate. To provide sufficient sugar cane plantations large tracts of forested land have been cleared and this could be ecologically disastrous. Much of the pressure in Brazil to produce alcohol is politically driven and politics may also influence the fortunes of the acetone–butanol fermentation. The political embargo on the export of oil to South Africa may trigger redevelopment of the acetone–butanol fermentation there to provide the raw materials for its manufacturing industry.

One advantage of the solvent fermentations over SCP production is that a less sophisticated facility is required to produce them. This certainly is true when one compares the processes used for acetone–butanol fermentation (see p. 5) and brewing with those required to grow microorganisms aseptically on methane or methanol. Therefore it is not inconceivable that suitable fermentation plants could be established in the developing countries.

Table 6.5 The origins of the different kinds of alcoholic beverages.

Alcoholic beverage	Origin
Non-distilled	
Beer	On germination, starch in barley grains is converted to sugar which is extracted by boiling in water to produce wort and this is fermented
Cider	Fermentation of apple juice
Wine	Fermentation of grape juice
Sake	Starch in steamed rice is hydrolysed with *Aspergillus oryzae* and the sugars released are fermented with yeast
Distilled	
Whisky (Scotch)	Distillation of alcohol produced from barley
Whiskey — Irish	Pot still whiskey produced from alcohol derived from a mixture of barley, wheat and rye. Grain whiskey produced from alcohol derived from maize
— Rye	Produced from alcohol derived from rye
— Bourbon	Produced from alcohol derived from maize
Rum	Distillation of fermented molasses, a by-product of sugar cane refining
Vodka	Distillation of alcohol produced from any non-grain carbohydrate source, e.g. potatoes
Gin	Distillation of alcohol derived from maize or rye and redistillation in presence of herbs and juniper berries
Tequila	Distillation of fermented extracts of Mexican cactus

ESSENTIAL PRIMARY METABOLITES

Essential primary metabolites are those compounds which microorganisms synthesize because they are essential for cell growth. A number of these compounds are of economic importance (Table 6.6). Microorganisms do not naturally overproduce these compounds because if they did so they would be at a competitive disadvantage relative to strains producing only enough for growth. This is because

Table 6.6 Economically important, essential primary metabolites produced by fermentation.

Metabolite	Use
Amino acids	
L-Glutamate	Flavour enhancer
L-Threonine	Feed supplement
L-Lysine	Feed supplement
L-Phenylalanine	Manufacture of Aspartame (artificial sweetener)
L-Tryptophan	Feed supplement
Vitamins	
Riboflavin (vitamin B_2)	Food supplement
Vitamin B_{12}	Food supplement and feed additive
Nucleosides	
Inosine-5′-monophosphate	Flavour enhancer
Guanosine-5′-monophosphate	Flavour enhancer
Pigments	
β-carotene	Precursor of vitamin A

overproduction wastes sources of carbon and energy. To obtain overproducing strains it is necessary to subvert the cellular regulatory mechanisms. Metabolic regulation is achieved in two ways: by *feedback inhibition* of the activity of key enzymes by the end-product of a biosynthetic pathway and *repression* of the synthesis of the enzymes in a pathway by the end-product. Overproduction can also be facilitated by eliminating competing biochemical reactions. These principles can best be explained by reference to a specific example, that of threonine overproduction.

A diagrammatic representation of threonine biosynthesis is shown in Fig. 6.6. Starting with a wild-type strain of *E. coli* that produces no excess threonine, a threonine-overproducer was obtained in three steps.

1 A mutant excreting 1.9 g/l was obtained by isolating a mutant resistant to the amino acid analogue α-amino, β-hydroxy valerate. In such mutants the enzyme homoserine dehydrogenase is no longer sensitive to feedback inhibition by threonine.

2 A mutant unable to convert threonine to isoleucine was selected following mutagenesis of the organism isolated in step 1 above. This mutant has two important properties. First, threonine is not converted to other compounds which reduce the overall yield. Second, the low internal concentration of isoleucine is insufficient to repress the synthesis of the enzymes converting homoserine to threonine and isoleucine. The threonine yield at this stage was 4.7 g/l.

3 The strain from step 2 was further mutated such that it now had a requirement for methionine. This raised the threonine yield to 6 g/l. Clearly, before such strains in steps 2 and 3 will grow, small amounts of methionine and isoleucine need to be added to the culture medium.

Another way to circumvent the regulatory controls of the cell is to alter the permeability of the

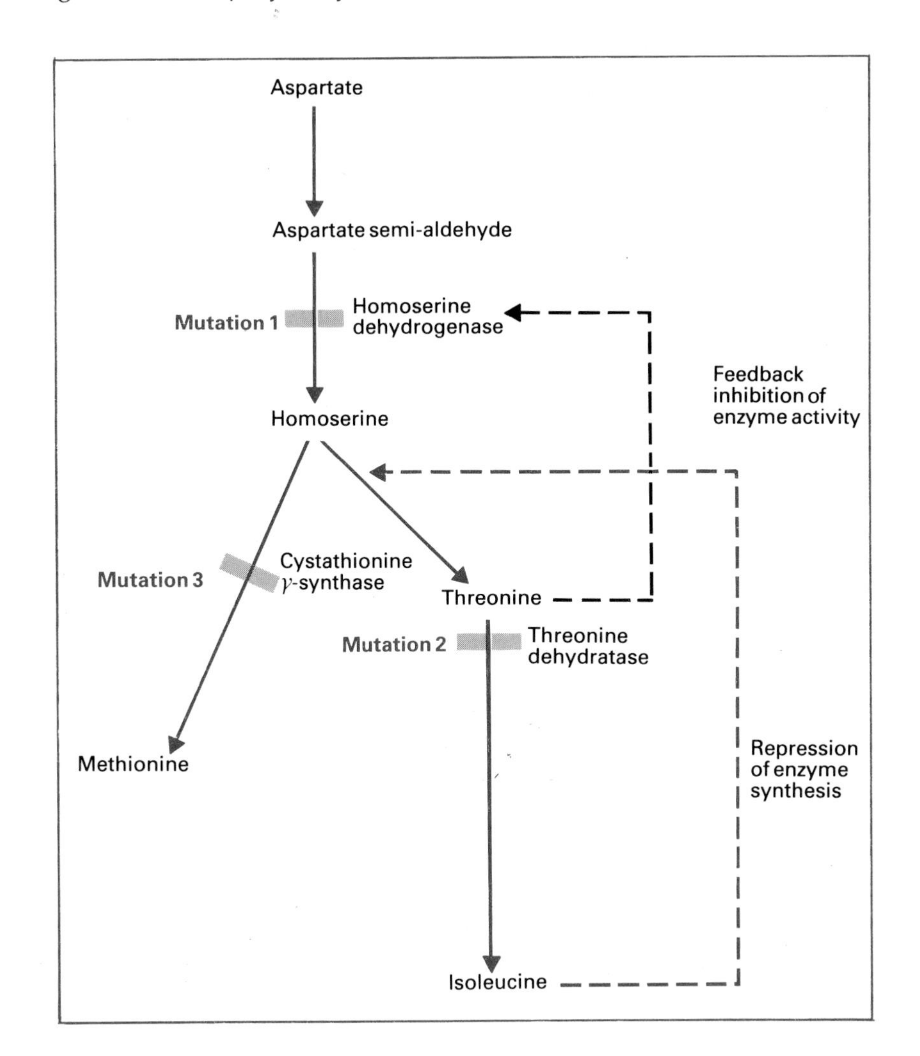

Fig. 6.6 Genetic manipulation of amino acid biosynthetic pathways leading to overproduction of threonine. Mutation 1 renders homoserine dehydrogenase insensitive to feedback inhibition by threonine (broken black arrow). Mutations 2 and 3 inactivate the enzymes indicated. See text for details.

cytoplasmic membrane. This enables the desired product to leak from the cell thereby reducing the intracellular concentration. In *Corynebacterium glutamicum,* which yields over 90 g glutamate/l, membrane permeability can be altered by biotin deficiency since biotin is essential for fatty acid synthesis and thus membrane integrity.

The removal of both the normal cellular regulatory controls and competing biochemical pathways can be achieved readily by conventional genetic techniques. Recombinant DNA technology also has a role to play; for example, one enzymic step may be rate-limiting thus reducing the metabolic flux through the pathway. By simply cloning the relevant gene such that there are multiple copies inside the cell the enzyme level can be amplified. However, there is an even better use for gene manipulation: clone the genes for all the enzymes in a biosynthetic pathway such that their synthesis is under the control of a single, well-characterized promoter, e.g. the *lac* promoter (see p. 14). Not only is the normal cellular repression of enzyme synthesis eliminated, synthesis of the enzymes can be controlled by the fermentation technologist. Where the producing organism is cultured in a series of fermenters of increasing size (see Fig. 5.7), synthesis can be repressed until the final production vessel is reached. In this way there is less problem of overgrowth of energetically fitter, non-producing mutants.

SECONDARY METABOLITES

Secondary metabolites are molecules synthesized by microorganisms late in the growth cycle (Fig. 6.7). They are not required for growth and their real function is not known. Since many of them, e.g. antibiotics, are inhibitory to other organisms they may impart an ecological advantage on the producing organism. The best known of the secondary metabolites are the antibiotics; others include mycotoxins and pigments. Over 2500 antibiotics have been described of which the majority are produced by actinomycetes. The diversity of molecular structures is impressive (Fig. 6.8) and many of them are produced as mixtures of related compounds. Thus there are over 20 penicillins, 20 actinomycins, 10 polymyxins, 10 bacitracins, 3 neomycins, etc.

Unlike primary metabolites such as amino acids, secondary metabolites as products were developed

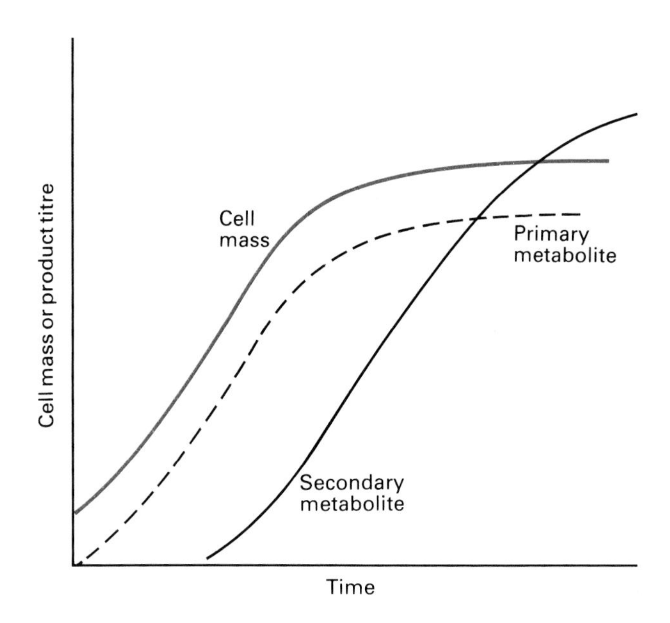

Fig. 6.7 Kinetics of formation of primary and secondary metabolites in relation to cell growth.

at a time when little was known about their pathways of biosynthesis. Even today details are unclear for many of them. Consequently two approaches have been used to obtain enhanced yield of product. One of these is the random mutation and selection procedure described earlier (see p. 6) and an example is shown in Fig. 6.9. The other involves screening of hundreds of culture medium additives as possible precursors of the desired product. Occasionally a precursor is identified that increases production of the secondary metabolite; for example, α-aminoadipate addition stimulates penicillin biosynthesis. Alternatively, the precursor may direct the formation of one specific desirable product ('directed biosynthesis'). Thus phenylacetic acid addition promotes the formation of benzylpenicillin over other penicillins. Both approaches have been very successful but both are labour intensive. Occasionally they can be combined. For example, where addition of an amino acid has been shown to be stimulatory then mutational efforts should be focused on removing regulatory controls on biosynthesis of that amino acid (as described in the previous section).

Development of superior antibiotic-producing strains has been hampered in two ways. First, the biosynthetic pathway often is not known making it difficult to apply selective mutation techniques. Second, the regulatory controls on synthesis are not

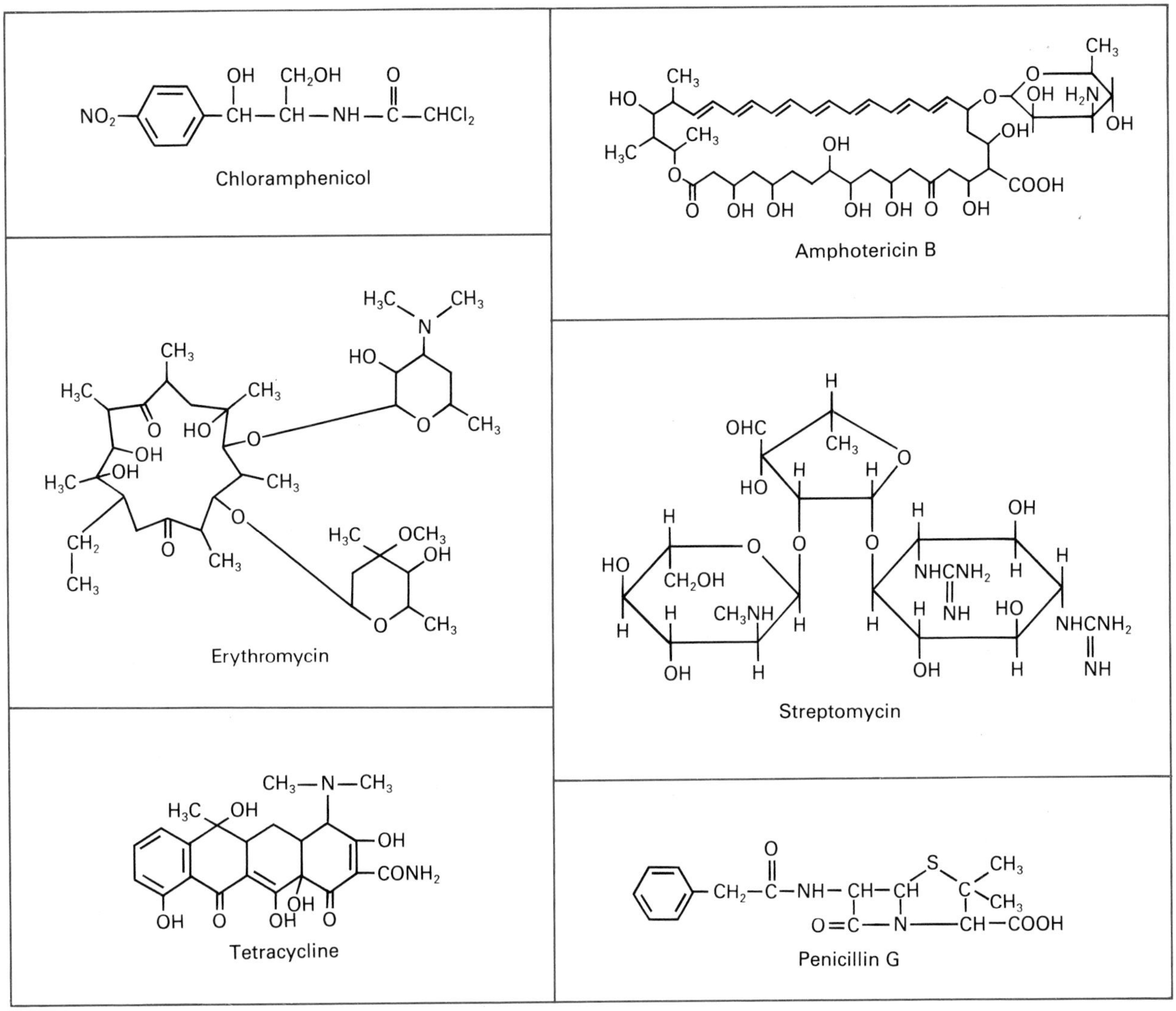

Chloramphenicol

Amphotericin B

Erythromycin

Streptomycin

Tetracycline

Penicillin G

Fig. 6.8 Structures of some representative antibiotics.

known, particularly that mechanism which delays synthesis until the end of the cellular growth phase. This is where recombinant DNA technology can be extremely helpful. The genes controlling entire biosynthetic pathways can be cloned. Once cloned, the DNA can be sequenced and this permits identification of the number of genes involved and is a good indication of the number of biosynthetic steps. Since the protein products of the cloned genes are overexpressed their purification is facilitated. By deleting one or more of the genes in the pathway, intermediates will accumulate and these can be identified by conventional analytical procedures, e.g. mass spectrometry. In this way a picture of the biosynthetic pathway can be constructed. More important, the normal regulatory circuits can be eliminated and the cloned genes placed under the control of a known promoter thus facilitating fermentation development.

If genes involved in antibiotic synthesis are to be cloned, then methods for their selection are required. A number of different approaches have been developed. One method is exemplified by candicin biosynthesis. Here a key step is the conversion of chorismate by the enzyme PABA synthetase. Cloning of the gene for the synthetase was facilitated by the availability of a direct selection method — restoration of PABA-independence to a

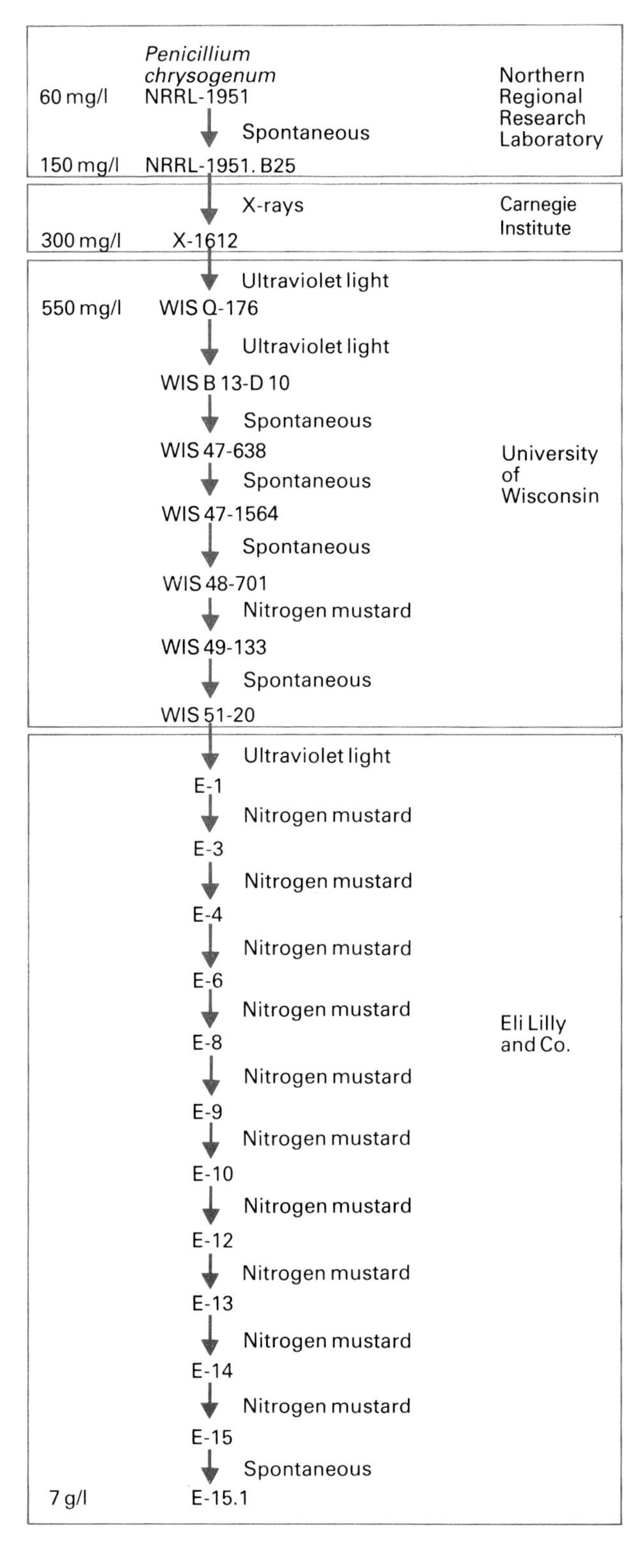

Fig. 6.9 Use of mutation and selection in the development of improved penicillin-producing strains of *P. chrysogenum* by mutation and selection.

strain requiring PABA for growth. A second approach is to isolate mutants which no longer produce the desired antibiotic and then identify DNA fragments which restore antibiotic synthesis. This is a useful approach if the antibiotic is coloured because it is easy to detect loss and restoration of antibiotic synthesis. The third procedure is to determine whether the genes for antibiotic synthesis are plasmid borne, as in the case of methylenomycin, for this facilitates their subsequent manipulation. The final method which can be used relies on the fact that an antibiotic-producing organism has to be resistant to that antibiotic otherwise it will kill itself. In some instances the genes specifying antibiotic-resistance are linked to those encoding antibiotic synthesis. For example, *Streptomyces lividans* does not produce any antibiotics related to erythromycin but when the *S. erythreus* gene for erythromycin-resistance was cloned and transferred to it, *S. lividans* became an erythromycin producer.

Just as novel proteins can be produced by recombinant DNA techniques (see Chapter 4), so too can novel antibiotics be produced. The *S. coelicolor* gene cluster encoding the biosynthesis of the isochromanequinone antibiotic actinorhodin has been cloned. When the cloned genes were introduced into a variety of other *Streptomyces* sp. producing different isochromanequinones at least three new antibiotics were detected. Clearly actinorhodin, or one of its precursors, is a novel metabolite in these other *Streptomyces* sp. and is subject to further or different enzymatic modifications. Yet another way of producing novel antibiotics is the provision of unusual substrates to enzymes involved in antibiotic biosynthesis. A key step in the synthesis of β-lactams (penicillin and cephalosporins) is the cyclization of a tripeptide precursor. When the purified enzyme is fed novel tripeptides novel β-lactams are produced.

BIOCONVERSIONS

Bioconversions (or biotransformations) are processes in which microorganisms convert a compound to a structurally related product. They comprise only one or a small number of enzymatic reactions as opposed to the multi-reaction sequences of fermentations. Although hundreds of different bioconversions have been described, they are only used commercially when conventional chemical approaches are too costly or difficult. For example, when stereoselective transformation is required, when there are no functional groups near the atom to be modified, or when only one of many identical functional groups in a molecule is to be modified. Some examples are shown in Fig. 6.10. The ultimate in specificity are the steroid bioconversions (Fig. 6.11). Their utility can best be illustrated by the synthesis of cortisone. The original, entirely chemi-

Progesterone — *Rhizopus nigricans* → II α -Hydroxyprogesterone

Phenylpyruvic acid (C_6H_5-$CH_2COCOOH$) — *Escherichia coli* → L-Phenylalanine (C_6H_5-CH_2CHNH_2COOH)

Cinnamic acid (C_6H_5-$CH{=}CHCOOH$) — *Rhodotorula* → L-Phenylalanine (C_6H_5-CH_2CHNH_2COOH)

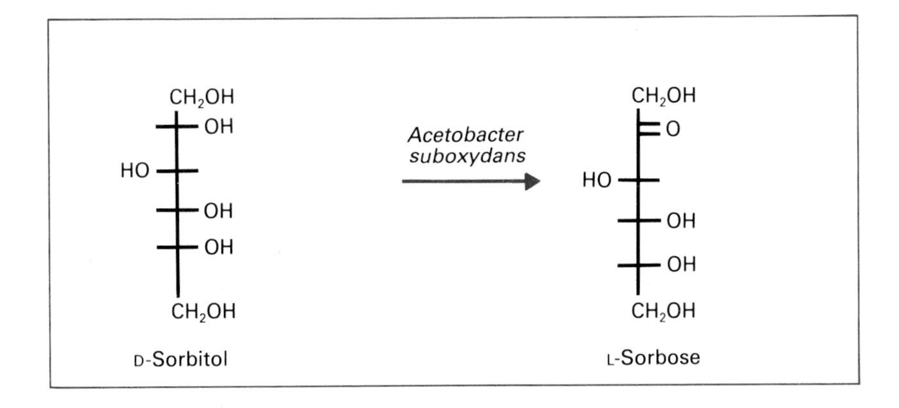

Fig. 6.10 Representative examples of commercial bioconversions.

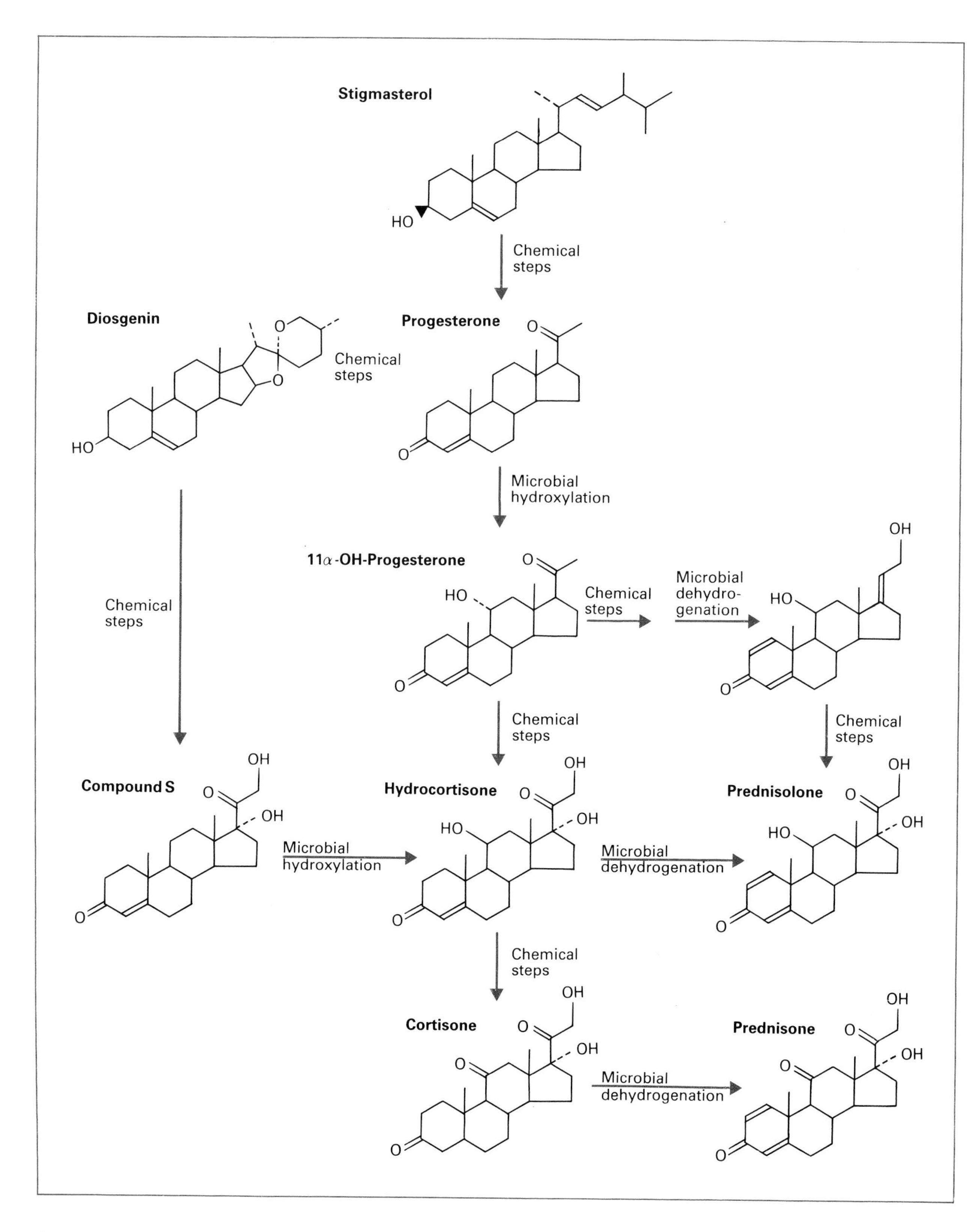

Fig. 6.11 The involvement of microorganisms in the production of therapeutically useful steroids. Diosgenin is extracted from the roots of the Mexican barbasco plant. Stigmasterol is extracted from soybeans.

cal synthetic route comprised 37 steps and the end-product cost \$200/g. The introduction of a microbial bioconversion reduced the number of steps to 11 and the cost dropped to \$6/g. Acetic acid bacteria are particularly useful for transformations of polyhydric alcohols, the most important being the conversion of sorbitol to sorbose (Fig. 6.10), an essential raw material for the chemical synthesis of vitamin C.

Although bioconversions can be conducted with growing cells, many are carried out with resting or killed cells or even spores. The advantages of using non-growing cells are threefold. First, very high substrate loadings can be used, e.g. tens of grams of substrate per litre, and such elevated substrate levels usually inhibit growth. Second, if washed cells are used, there will be no contaminating substances present. These two features make product isolation very easy. So too does the fact that substrate conversion efficiencies of 85–100% are readily achieved. Finally, many enzymes have pH and/or temperature optima far removed from those of the intact cell. In bioconversions these can be optimized to increase the reaction rate.

Another feature of bioconversions is that they are easy to scale-up since the only parameter of interest is the level of the enzyme mediating the transformation. This makes bioconversions obvious candidates for the application of recombinant DNA technology. Not only can the enzyme levels be increased in existing processes but processes which previously were uneconomic can be reconsidered. Thus the levels of aspartate aminotransferase in wild-type *E. coli* are too low to make this organism of much use for L-amino acid synthesis from α-keto acids (Fig. 6.10). By cloning the relevant gene, enzyme levels can be raised to over 20% of total soluble protein and bioconversion becomes feasible. Once the gene is cloned it is possible to use site-directed mutagenesis (see p. 35) to produce enzyme variants with enhanced thermostability or altered substrate profile. It is also possible to develop novel synthetic routes and the best example is provided by the manufacture of vitamin C. The conventional process starts with glucose and comprises one microbiological and four chemical steps (Fig. 6.12(a)). By cloning in *Erwinia* a gene from *Corynebacterium* the process can be simplified to a single microbiological and a single chemical step (Fig. 6.12(b)). Another variant of this concept of new synthetic routes is provided by the microbial synthesis of indigo. Cloning of a single *Pseudomonas* gene,

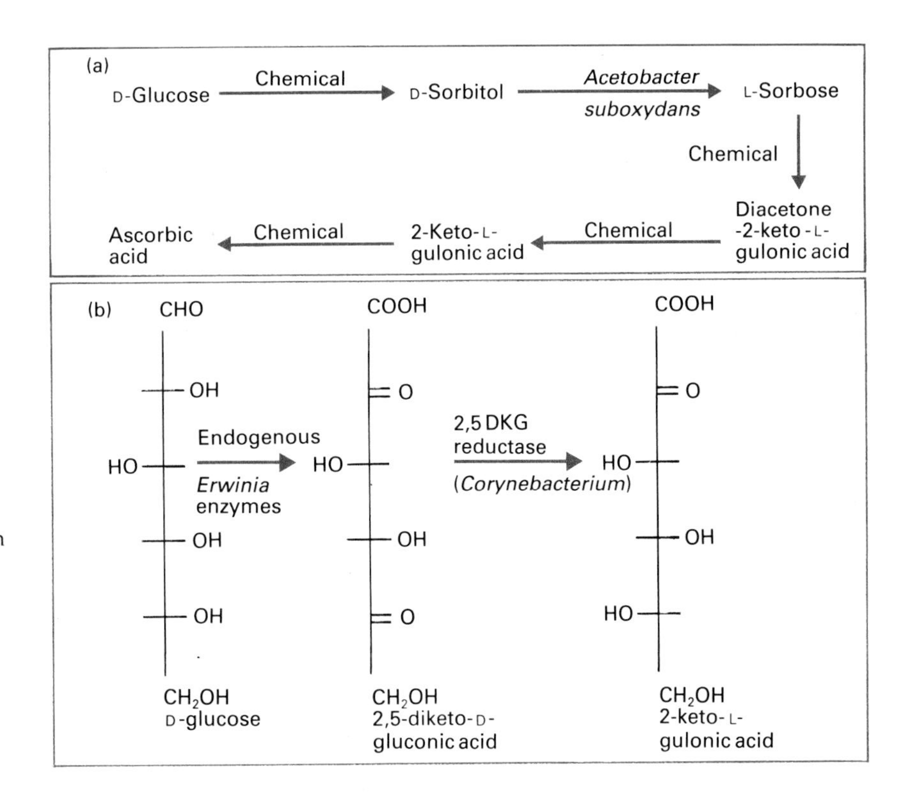

Fig. 6.12 Simplified route to vitamin C (ascorbic acid) developed by cloning in *Erwinia* the *Corynebacterium* gene for 2,5 diketogluconic acid reductase. (a) Classical route to vitamin C. (b) The simplified route to 2-ketogulonic acid, the immediate precursor of vitamin C.

that encoding naphthalene dioxygenase, resulted in the generation of an *E. coli* strain able to synthesize indigo from tryptophan (Fig. 6.13).

Production of high molecular weight compounds

Microorganisms are used to produce two groups of high molecular weight compounds (macromolecules): polysaccharides and proteins. The commercial production of both has increased dramatically in the last 5–10 years but for different reasons. Recombinant DNA technology has impacted on protein production because now it is possible to overproduce proteins which normally are synthesized in trace amounts and consequently have been too costly and difficult to isolate. By contrast, polysaccharides traditionally have been isolated from plants, particularly seaweeds. This has disadvantages including high labour costs for harvesting and extraction, limited availability of the sources, a supply that can be affected adversely by climate, and seasonal variation in polysaccharide composition. Microorganisms can provide a constant and reliable supply of a polymer with relatively uniform chemical and physical properties.

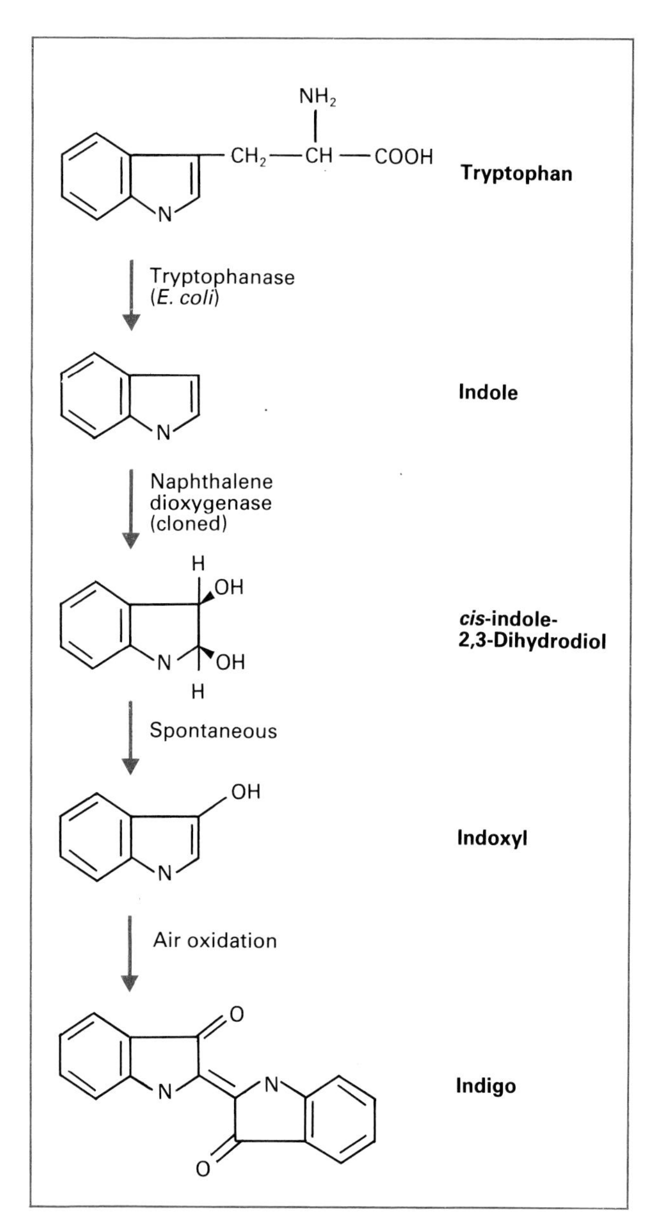

Fig. 6.13 The production of indigo by *E. coli* carrying a cloned naphthalene dioxygenase gene. Tryptophan in the medium is converted to indole by endogenous tryptophanase.

MICROBIAL POLYSACCHARIDES

The industrial value of polysaccharides lies in their capacity for altering the rheological properties of aqueous solutions, either through gelling or by the alteration of their flow characteristics. Many polysaccharides are thixotropic, i.e. the solutions are characterized by high viscosity at low stress and decreased viscosity when stress is applied. Such characteristics make them particularly useful in a whole variety of industries (Table 6.7). Pullulan has potential applications in the cosmetic and food industries but its greatest use may be as a biodegradable plastic. The clear films and plastic containers currently in use are made from petroleum-based products which are not biodegradable.

The greatest single potential market for polysaccharides is the oil industry. Conventional oil extraction technologies can recover only about 50% of the world's subterranean oil reserves, the balance is either trapped in rock or too viscous to pump. The application of microorganisms, or their products, to aid in the recovery of this oil is known as *microbial-enhanced oil recovery* (MEOR). For example, biological compounds that could be injected into wells include surfactants and viscosity decreasers to release trapped oil, and viscosity enhancers to push oil out of crevices. Xanthan gum can be used as a viscosity enhancer and emulsan as a surfactant. Like many biologically produced polymers, emulsan exhibits a specificity that generally is not observed in chemically synthesized materials; the

Table 6.7 Commercially available microbial polysaccharides and their uses.

Polysaccharide	Producing organism	Uses
Xanthan gum	*Xanthomonas campestris*	1 Food additive for stabilizing liquid suspensions and gelling soft foods, e.g. ice cream, cheese spreads 2 Lubrication in, for example, toothpaste preparations 3 Enhanced oil recovery
Gellan	*Pseudomonas* sp.	Solidification of food products
Emulsan	*Acinetobacter calcoaceticus* *Arthrobacter*	1 Cleaning oil spills 2 Enhanced oil recovery
Pullulan	*Aureobasidium pullulans*	1 Biodegradable material for food coating and packaging
Dextrans	*Leuconostoc mesenteroides*	1 Blood expander 2 Adsorbents for pharmaceutical preparations

emulsifying activity of emulsan is substrate-specific, acting only on hydrocarbons that have both aliphatic and cyclic components.

Polysaccharides are found outside the microbial cell wall and membrane, hence the use of the term *exopolysaccharide*. Their synthesis is favoured by an excess of carbon substrate in the growth medium, particularly if nitrogen is limiting. Although they are mostly produced in batch culture, chemostat culture could be used. In this instance, as with biomass production, genetic instability is not a problem. Furthermore, one can vary the composition of the polysaccharide by varying the limiting nutrient; for example the polysaccharide may be composed mostly of neutral sugars if nitrogen is limiting but when metal ions are limiting there can be a high proportion of sugar acids (e.g. glucuronic acid).

With many microorganisms there are probably over 100 enzymatic steps directly or indirectly involved in polysaccharide biosynthesis. This is because microbes have the ability to produce several polysaccharides of different sugar composition and sequence. Additionally, subtle chemical alterations to one or more sugar residues can have dramatic effects on the physicochemical properties of the molecule. Thus for the introduction of process and product improvements it is likely that classical genetic techniques initially will be more useful than recombinant DNA technology.

PRODUCTION OF PROTEINS

The microbial production of proteins is exploited commercially in two ways: the production of bulk enzymes and the production of much smaller quantities of proteins of therapeutic value. Bulk enzymes have a variety of applications (Table 6.8). Their manufacture is a well-established business and by the end of the decade the market will be about 100 000 tonnes with a value of $1 billion. Fewer than 20 enzymes comprise the large part of this market and only a proportion of them are produced by microorganisms. Because enzymes are direct gene products, theoretically they are good candidates for improved production through recombinant DNA technology. However, many of the production processes already are high yielding, e.g. in excess of 20 g/l, and significant increases may not be possible. The use of gene manipulation techniques to modify proteins, as discussed in Chapter 4, may be more important because they could yield enzymes with improved properties. For example, a lowering of the pH optimum of glucose isomerase used in the production of high-fructose corn syrups would reduce the product discoloration ('browning reaction' or 'caramelization') that occurs with the normal enzyme. There is also the potential for increased thermostability. Cloning also could be used to provide alternative sources of enzymes which are in short supply. One such enzyme is rennet (chymosin), which is obtained from the stomach of young

Table 6.8 Sources and applications of some microbial enzymes.

Enzyme	Source	Applications
α-amylase	*Aspergillus oryzae*	Preparation of glucose syrups
	Bacillus amyloliquefaciens	Removal of starch sizes
	Bacillus licheniformis	Liquefaction of brewing adjuncts
β-glucanase	*Aspergillus niger*	Liquefaction of brewing adjuncts
	Bacillus amyloliquefaciens	Improvement of malt for brewing
Glucoamylase	*Aspergillus niger*	Starch hydrolysis
	Rhizopus sp.	
Glucose isomerase	*Arthrobacter* sp.	High-fructose corn syrup
	Bacillus sp.	
Lactase	*Kluyveromyces* sp.	Removal of lactose from whey
Lipase	*Candida lipolytica*	Flavour development in cheese
Pectinase	*Aspergillus* sp.	Clarification of wines and fruit juices
Penicillin acylase	*Escherichia coli*	Preparation of 6-aminopenicillanic acid
Protease, acid	*Aspergillus* sp.	Calf rennet substitute
Protease, alkaline	*Aspergillus oryzae*	Detergent additive
	Bacillus sp.	Dehairing of hides
Protease, neutral	*Bacillus amyloliquefaciens*	Liquefaction of brewing adjuncts
	Bacillus thermoproteolyticus	
Pullulanase	*Klebsiella aerogenes*	Starch hydrolysis

calves. This enzyme is used widely in the cheese industry to facilitate curdling of the milk (see p. 65). Currently demand outstrips supply and a variety of microbial proteases ('microbial rennet') are used but these are of inferior quality since they generate unwanted flavours. The gene for chymosin has been cloned and commercial production will be realized soon.

One protein which is not an enzyme but which will be produced in bulk is bovine growth hormone. Administration of this hormone to cows significantly increases their milk yield. The hormone also has an effect on fish. Weekly injection of the hormone into young salmon results in a doubling in the rate of weight gain but it remains to be seen if this practice will be adopted by fish farms.

Perhaps the most glamorous aspect of recombinant DNA technology is its use to construct micro-organisms which synthesize human proteins with therapeutic potential. Many of these proteins are synthesized in trace amounts in the body, particularly if they are hormones, and before the advent of gene manipulation many had not been purified or even isolated in crude form. A list of some of those currently being investigated is given in Table 6.9. So far only three are being marketed: human insulin, human growth hormone and α_2-interferon. Many more are undergoing clinical trials. Protein engineering has already been used to generate novel proteins which potentially have improved therapeutic properties (see p.40). Attempts have also been made to use genetic engineering to facilitate

Table 6.9 Some human proteins with therapeutic potential.

Protein	Application
Interferons	Treatment of virus infections and cancer
Human growth hormone	Pituitary dwarfism
Insulin	Diabetes
Tissue plasminogen activator	Thrombosis
Urokinase	Thrombosis
Epidermal growth factor (urogastrone)	Wound healing
Interleukin-2	Cancer therapy
Relaxin	Facilitation of childbirth
α_1-antitrypsin	Emphysema
Tumour necrosis factor	Cancer therapy
Erythropoietin	Treatment of anaemia
Lung surfactant protein	Treatment of respiratory distress syndrome

vaccine production. Thus the genes for the major subunits of a number of viruses, e.g. foot and mouth disease virus, have been cloned in *E. coli* but the results have been disappointing: the peptides thus produced have been poorly immunogenic. One notable exception is a hepatitis B vaccine prepared in yeast and which has received regulatory approval from the US Food and Drug Administration. Approaches using recombinant animal cells are more promising and are discussed on page 111.

Regardless of whether a bulk enzyme or a therapeutic protein is being produced, downstream processing will involve protein purification. Standard biochemical techniques are used, e.g. ion exchange chromatography, and protein engineering can be used to facilitate purification, as the example of polyarginine-tailing shows (see p. 43). However, there is a fundamental difference in the purity requirements of the two classes of protein. With bulk proteins the major criterion is activity: the greater the activity of the protein the more useful it will be. Purity is much less important and if some of the enzyme is denatured during processing it is of little consequence provided the specific activity does not fall below a minimally acceptable level. By contrast, most of the therapeutically useful proteins will be administered parenterally and they must be of the highest quality (Table 6.10). When foreign proteins are synthesized in microorganisms the ends of the molecules can be 'nibbled' by proteases and such partially degraded molecules must be removed for they could become immunogenic if therapy were prolonged. An added complication is that, in *E. coli* at least, many foreign proteins are deposited in inclusion bodies (Fig. 6.14). The protein has to be extracted from these inclusions using denaturing agents and the protein subsequently re-folded. This is a process akin to recreating native egg white from a meringue! If significant protein loss is not to occur during downstream processing, care must be taken to exclude microorganisms at all stages of purification. Finally, the overall process has to be conducted according to the code of *good manufacturing practice* (GMP) (see Box, p. 82).

Table 6.10 Specification for therapeutic proteins to be administered parenterally.

Greater than 95% purity
Microheterogeneity below specified level
Contaminating DNA less than 10 pg/dose
Endotoxin below specified level
Toxic chemicals (used in purification) below specified level
Specific activity above a minimum level and preferably maximum possible
Absence of microorganisms

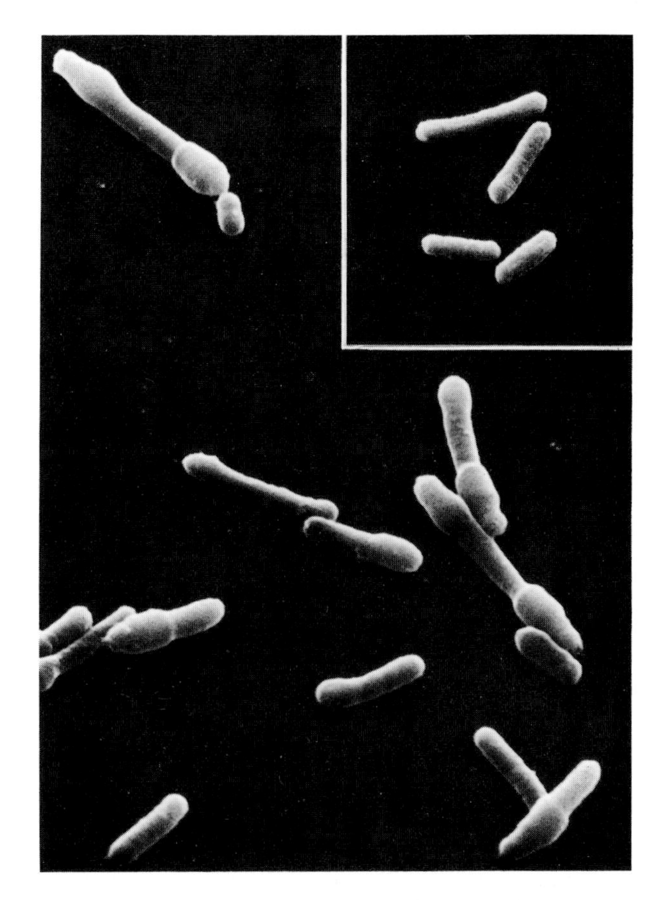

Fig. 6.14 Protein inclusions in *E. coli* overproducing a recombinant protein. The inset shows normal *E. coli* cells. Photograph reproduced from *Science* courtesy of Dr. D.C. Williams (Eli Lilly & Co.) and the American Association for the Advancement of Science.

Processes dependent on general microbial metabolism

The one microbiological process that most affects the quality of modern life, at least in the developed countries, is the disposal of industrial and domestic effluent. Without sewage treatment plants we would be awash with animal and human excreta and other noxious substances, disease would be more prevalent and we would have difficulty in obtaining water of suitable quality for human or industrial use. The disposal of effluent is a simple process and is shown diagrammatically in Fig. 6.15. Successful

The code of good manufacturing practice (GMP)

GMP guidelines can and should be applied in any manufacturing operation. As far as biotechnology is concerned these guidelines are most appropriate to the production of therapeutic proteins. According to the UK guide to GMP, 'there should be a comprehensive system, so designed, documented, implemented and controlled, and so furnished with personnel, equipment and other resources as to provide assurance that products will be consistently of a quality appropriate to their intended use.' The basic requirements of GMP are that:

1 all manufacturing processes should be clearly defined, and known to be capable of achieving the desired ends;

2 all necessary facilities should be provided, including:

(a) appropriately trained personnel;
(b) adequate premises and space;
(c) suitable equipment and services;
(d) correct materials, containers and labels;
(e) approved procedures (including cleaning procedures);
(f) suitable storage and transport;

3 procedures are to be written in instructional form, in clear and unambiguous language, and are to apply specifically to the facilities provided;

4 operators should be trained to carry out the procedures correctly;

5 records are to be kept during manufacture (including packaging) which demonstrate that all the steps required by the defined procedures are in fact taken and that the quantity and quality produced are those expected;

6 records of manufacture and distribution which enable the complete history of a batch to be traced should be retained in a legible and accessible form;

7 a system must be available for recall from sale or supply any batch or product should that become necessary.

GMP is often confused with *quality assurance* and *quality control*. Quality assurance encompasses all the organized arrangements made with the object of ensuring that the products will be fit for their intended use. It includes GMP and other factors such as original product design, product development, storage and distribution of the final product. Quality control is that part of GMP which is concerned with sampling and testing to ensure that the product meets specification.

operation relies on the fact that sewage contains many different organic chemicals which can serve as substrates for microbial growth. The basic principle is the conversion of organic wastes to microbial cells and for maximum efficiency has to be an aerobic process. In the activated sludge process the sewage is vigorously agitated in large, open tanks. In the trickling filter process sewage is allowed to trickle down columns of coarse gravel through which air can permeate. Both processes are operated continuously. A number of newer designs are being introduced which are more efficient in terms of land use and energy consumption.

The digestion process used to treat the sludge resulting from the primary and secondary treatments is an anaerobic process. Its purpose is threefold: to reduce the volume of solids requiring disposal; to reduce the odour; and to reduce the number of pathogenic organisms. In some instances the methane produced during anaerobic digestion can be recovered and used to generate electricity. Ultimately the digested sludge is used for landfill or as a general-purpose fertilizer.

The objective of waste disposal systems is to degrade all the organic molecules in the effluent. However, during the past quarter of a century a number of man-made chemicals resistant to biodegradation have been released into the environment. Plastics and nylon are good examples and their persistence is due to the fact that they do not exist naturally and, having unusual chemical linkages, microorganisms do not exist which can degrade them. Of particular significance are compounds such as chlorinated hydrocarbons because, unlike plastics, they are not biologically inert but are cytotoxic, mutagenic and carcinogenic. Examples of

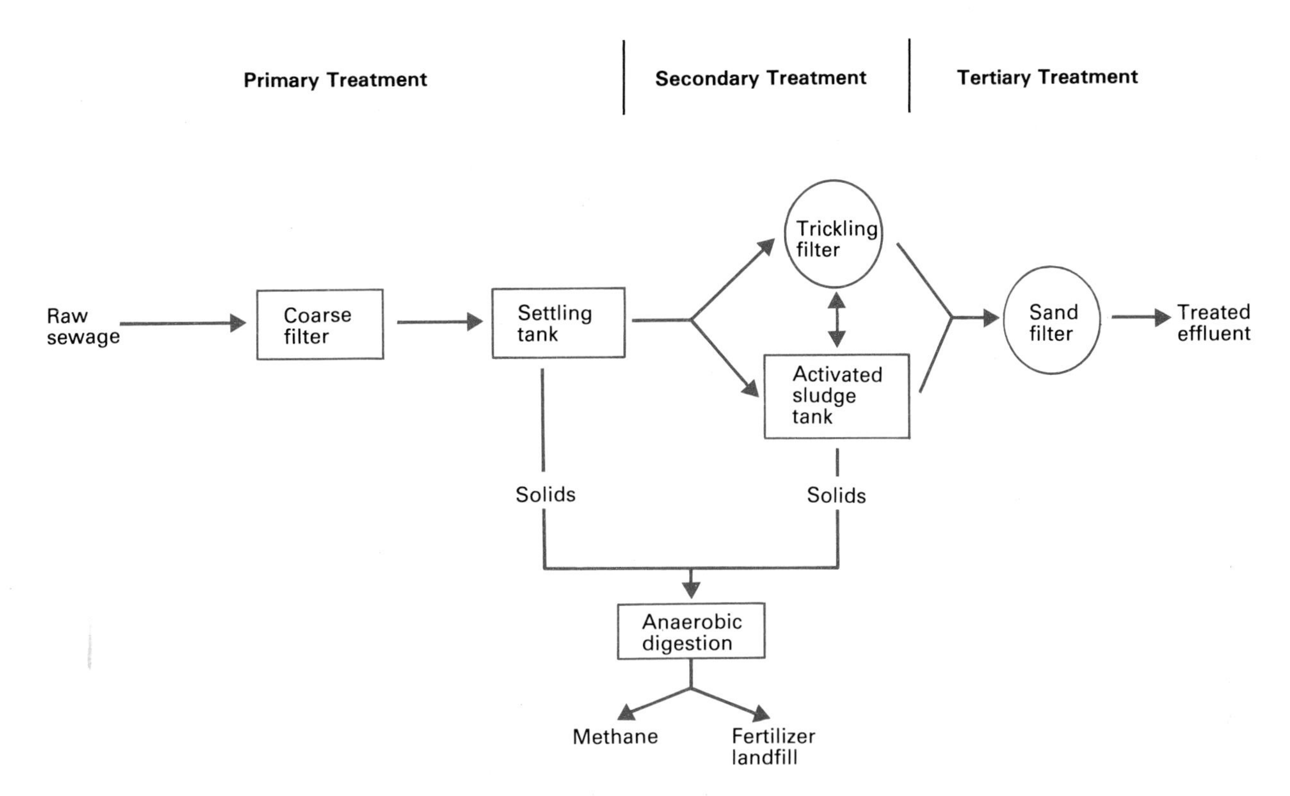

Fig. 6.15 Schematic representation of a sewage treatment works. In some plants effluent passes through both an activated sludge tank and a trickling filter.

such toxic chemicals are DDT and the herbicide 2,4,5-trichlorophenoxyacetic acid (2,4,5-T, Agent Orange).

Bacterial plasmids have been identified which carry genes specifying the degradation of chlorinated aromatic compounds such as 3-chlorobenzoate. Cells carrying these plasmids do not metabolize 4-chlorobenzoate unless a second plasmid encoding toluene and xylene decomposition is also present. By combining plasmids in this way it is possible greatly to expand the range of substrates which microbes can metabolize. After a range of microorganisms carrying different plasmids encoding degradation of various aromatic compounds were incubated with 2,4,5-T for 8–10 months an organism capable of growing on 2,4,5-T as sole carbon source was isolated. This organism contains a plasmid which has recruited genes from a variety of different plasmids. Although this genetic rearrangement occurred spontaneously, the process could be speeded up by the use of recombinant DNA technology. Based on a knowledge of the biochemistry of catabolism of aromatic compounds it should be possible to identify those genes which need to be united.

MICROBIOLOGICAL MINING

Microorganisms have been used in mineral leaching and metal concentration for many years. Indeed, more than 10% of the copper produced by the US is leached from ores by microorganisms. The microorganisms used are found naturally associated with ores; the ores are not deliberately inoculated. To date, the use of microorganisms in the mining industry has received relatively little attention because of the ease of mining high grade ores. As high grade ores are depleted and energy costs soar microbial leaching becomes more attractive. Currently leaching is restricted to copper and uranium ores but the commonest organism involved, *Thiobacillus ferrooxidans*, can also effect the solubilization of cobalt, nickel, zinc and lead.

The leaching of metals from ores can occur directly or indirectly. In the direct attack on metal sulphides the reaction which occurs is:

$$MS + 2O_2 \rightarrow MSO_4$$

where M can be Zn, Pb, Cu, Co, Ni, etc. In the indirect attack, pyrites (FeS_2) is oxidized to ferric sulphate and sulphuric acid and the ferric sulphate oxidizes the metal sulphide, e.g.:

$$4\ FeS_2 + 15O_2 + 2H_2O \rightarrow 2\ Fe_2(SO_4)_3 + 2\ H_2SO_4$$
$$CuS + Fe_2(SO_4)_3 \rightarrow CuSO_4 + 2FeSO_4 + S^0.$$

Whether direct or indirect attack occurs, the result is the same. The metal sulphate-containing solutions are collected and the metals removed by chemical precipitation.

Ways in which biotechnology could enhance the efficiency of microbial leaching are shown in Table 6.11. Of those listed, the easiest to achieve would be increasing the resistance to heavy metals. Unfortunately, *T. ferrooxidans* is a difficult organism to cultivate and thus not particularly amenable to genetic manipulation. But, where there is a will, there is a way!

Table 6.11 Desirable changes in *Thiobacillus ferrooxidans* strains used in microbiological mining.

1 An enhancement of the rate of regeneration of the ferric ion
2 Greater tolerance to acidic conditions
3 Greater tolerance to saline conditions
4 Increased resistance to metal ions such as Ag, Hg, Cd
5 Increased ability to withstand high temperatures

Further reading

GENERAL

Bull A.T., Ellwood D.C. & Ratledge C. (1979) *Microbial Technology: Current State, Future Prospects.* Symposium 29 of the Society for General Microbiology. Cambridge University Press, Cambridge.

Peppler H.J. & Perlman D. (1979) *Microbial Technology*, vols I and II, 2nd edn. Academic Press, London.

Rehm H-J. & Reed G. (1983 onwards) *Biotechnology, a Comprehensive Treatise*, vols 2–7. Springer-Verlag, Basle.

Each of these treatises contains one or more specific articles dealing at length with topics covered in this chapter.

SPECIFIC

Anderson S., Marks C.B., Lazarus R., Miller J., Stafford K., Seymour J., Light D., Rastetter W. & Estell D. (1985) Production of 2-keto-L-gulonate, an intermediate in L-ascorbate synthesis, by a genetically modified *Erwinia herbicola*. *Science* **230**, 144–9.

Armstrong D.W. & Yamazaki H. (1986) Natural flavours production: a biotechnological approach. *Trends in Biotechnology* **4**, 264–8.

Aronson A.I., Beckmann W. & Dunn P. (1986) *Bacillus thuringiensis* and related insect pathogens. *Microbiological Reviews* **50**, 1–24.

Bonnerjea J., Oh S., Hoare M. & Dunnill P. (1986) Protein purification: the right step at the right time. *Bio/technology* **4**, 954–8.

Dalbøge H., Dahl H-H.M., Pedersen J., Hansen J.W. & Christensen T. (1987) A novel enzymatic method for production of authentic hGH from an *Escherichia coli* produced hGH-precursor. *Bio/technology* **5**, 161–4.

Green M.L., Angal S., Lowe P.A. & Marston F.A.O. (1985) Cheddar cheesemaking with recombinant calf chymosin synthesized in *Escherichia coli*. *Journal of Dairy Research* **52**, 281–6.

Klausner A. (1984) Microbial insect control: using bugs to kill bugs. *Bio/technology* **2**, 408–19.

Miyagawa K., Kimura H., Nakahama K., Kikuchi M., Doi M., Akiyama S. & Nakao Y. (1986) Cloning of the *Bacillus subtilis* IMP dehydrogenase gene and its application to increased production of guanosine. *Bio/technology* **4**, 225–8.

Primrose S.B. (1986) The application of genetically-engineered microorganisms in the production of drugs. *Journal of Applied Bacteriology* **61**, 99–116.

Scolnick E.M., McLean A.A., West D.J., McAleer W.J., Miller W.J. & Buynak E.B. (1984) Clinical evaluation in healthy adults of a Hepatitis B vaccine made by recombinant DNA. *Journal of the American Medical Association* **251**, 2812–15.

Sherwood M. (1984) The case of the money-hungry microbe. *Bio/technology* **2**, 606–9.

Stanzak R., Matsushima P., Baltz R.H. & Rao R.N. (1986) Cloning and expression in *Streptomyces lividans* of clustered erythromycin biosynthesis genes from *Streptomyces erythreus*. *Bio/technology* **4**, 229–32.

Vasey R.B. & Powell K.A. (1984) Single-cell protein. *Biotechnology and Genetic Engineering Reviews* **2**, 285–311.

7/Cell and Enzyme Immobilization

Microorganisms act as a catalyst in any fermentation where a product other than biomass is obtained. When the culture medium is removed from the fermenter the catalyst is either destroyed by subsequent downstream processing or discarded. It is possible to harvest cells and reuse them but there are a number of practical problems. Thus enzyme activity is lost unless there is a further period of cell growth and contamination is a major problem. One solution is to immobilize the microorganisms or even the appropriate enzyme.

The use of immobilized cells is not a new idea. Since the early eighteenth century they have been used in the production of vinegar and since the early part of this century for disposal of sewage effluent. In the vinegar process alcohol is trickled down a column of beechwood shavings which are coated with a film of acetic acid bacteria that carry out the reaction:

$$CH_3CH_2OH \longrightarrow CH_3CHO \longrightarrow CH_3COOH.$$

$$NAD^+ \curvearrowright NADH + H^+ \qquad NAD^+ \curvearrowright NADH + H^+$$

A similar technique is used for sewage effluent disposal except that sand and gravel act as the microbial support and the desired end-result is conversion of all the organic materials in the effluent to cells, water, carbon dioxide and other gases. In both instances the microbial film forms naturally and is held in place by a combination of microbially produced exopolysaccharides ('slime') and physical forces. Since growth of the organisms occurs, enzyme decay is not a problem. There is also a continual release of aggregates of microorganisms ('flocs').

Although these processes for vinegar manufacture and sewage disposal are perfectly satisfactory, it would be preferable if in other processes using immobilized cells the effluent stream were free of microorganisms. In order to achieve this the cells must be firmly bonded to the support matrix or trapped in some other way and cell growth must not occur. The latter demands that the requisite enzymes have long half-lives. As it turns out, these criteria can be met.

IMMOBILIZATION METHODS

The methods of immobilization available are equally applicable to cells and enzymes. Three categories of method can be discerned: physical binding, cross-linking and entrapment (Fig. 7.1). In practice, combinations of methods may be used. Physical binding is the oldest immobilization technique but the least satisfactory. Cells or enzyme are mixed with adsorbent and then packed in a column. However, ease of adsorption also means ease of desorption and this often occurs following substrate addition. Cells immobilized in this way have a tendency to autolyse and if enzymes are immobilized activity is often partially or totally lost.

Enzymes and microbial cells can be immobilized by cross-linking them with bi- or multi-functional reagents such as glutaraldehyde or toluene diisocyanate. They also can be bonded to insoluble matrices using the same reagents. Although good retention of cells and enzymes is obtained with this method, it can be accompanied by extensive loss of enzyme activity.

The most extensively used method of cell immobilization involves entrapment in a polymer matrix. Matrices which have been employed include collagen, gelatin, agar, alginate, carrageenan, polyacrylamide, cellulose triacetate and polystyrene. For immobilization with acrylamide cells are mixed with acrylamide monomer, a polymerizing agent such as N,N′-methylene bisacrylamide, potassium persulphate to initiate polymerization and β-dimethylaminopropionitrile, which is an accelerator of polymerization. After 30–60 min at ambient temperatures a stiff gel forms and this can be granulated to a suitable size, packed in a column and washed with saline to remove residual chemicals. Carra-

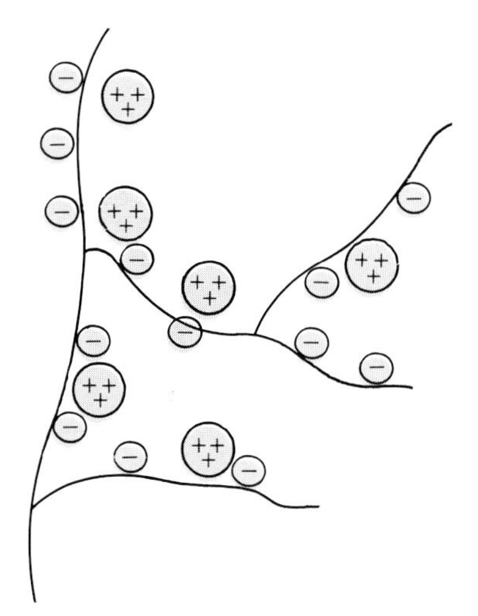

1 Physical association

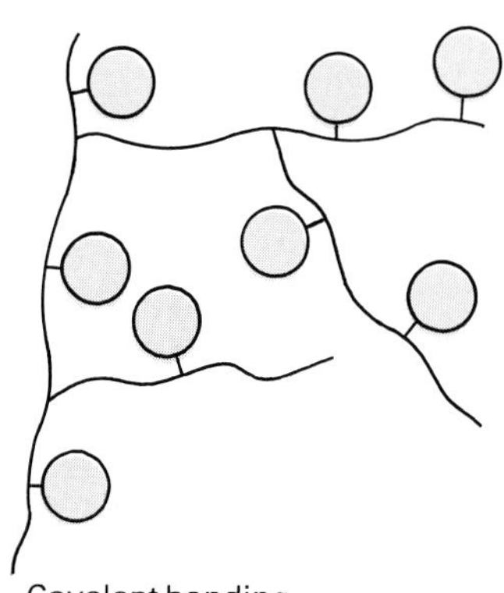

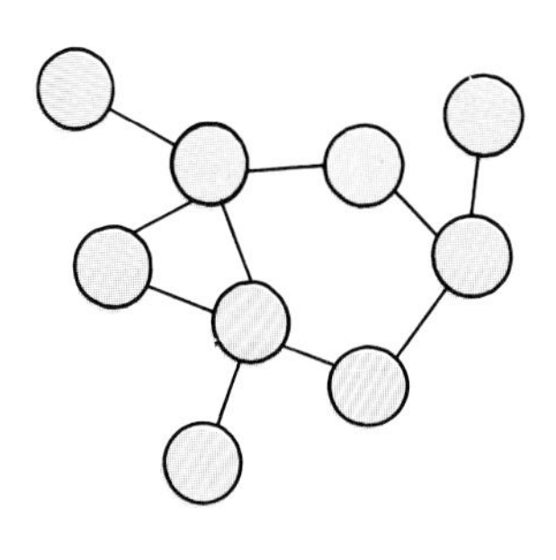

Covalent bonding

Copolymerization

2 Cross-linking

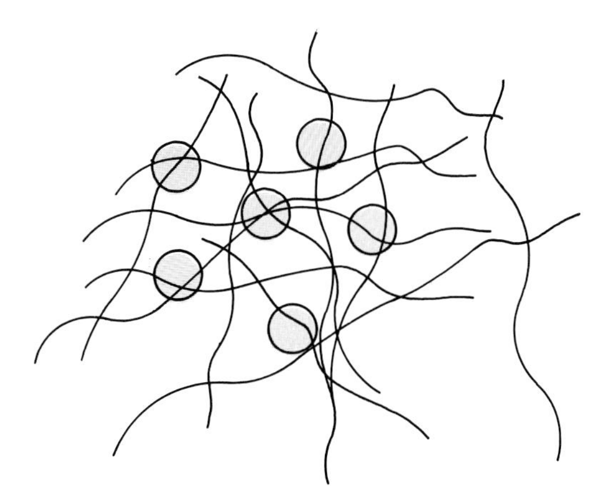

3 Entrapment

Fig. 7.1 The three basic methods of cell and enzyme immobilization. The tinted circles represent enzyme molecules or microbial cells.

geenan in some ways is an even more advantageous matrix for immobilization: the procedure used is much simpler and uses no noxious chemicals so that there is better retention of enzyme activity. Cells are suspended in saline at 45–50 °C and carrageenan is dissolved in saline at a similar temperature. The two solutions are mixed and cooled to 10 °C. To increase the gel strength it is soaked in cold potassium chloride. After this treatment the gel is granulated to a suitable size. If the operational stability of the support matrix is not satisfactory, then a hardening agent such as glutaraldehyde or hexamethylenediamine can be added.

CHOICE OF IMMOBILIZATION METHOD

The art of immobilization lies in knowing which method to choose. This, in turn, is governed by a number of factors, some of which will not be apparent until the procedure is tried. At least five factors are likely to be of paramount importance:

1 the operational stability of the immobilized catalyst;

2 the cost of the immobilized catalyst;

3 the activity and yield of the immobilized catalyst;

4 the regenerability of the catalyst;

5 the cost of the appropriate reactor configuration.

Initially the choice of method is empirical but subsequently the process must be thoroughly analysed and tested and considered on its merits.

EXAMPLE: IMMOBILIZATION OF L-AMINO ACID ACYLASE

There is a large market for L-amino acids as food and feed supplements. Some L-amino acids can be made efficiently by fermentation, others cannot and have to be prepared chemically. Although chemical synthesis often is easier and cheaper, the intermediate products are usually a mixture of the D- and L-isomers and resolution is necessary. One method of resolution is to acylate the amino acid (Fig. 7.2) and selectively hydrolyse the L-acyl derivative with the enzyme aminoacylase. The L-amino acid and the D-acyl amino acid can be separated easily by virtue of their differing solubilities. The D-acyl amino acid can be discarded or, better, racemized to the D,L-form and reused.

Industrial processes for L-amino acid production based on the use of soluble aminoacylase were introduced over 30 years ago. These processes suffered from two basic drawbacks: complicated separation of product and enzyme with concomitant low yields and high costs, and non reusability of the enzyme. To overcome these problems immobilization techniques were developed in Japan by Chibata and Tosa and in 1969 aminoacylase became the first immobilized enzyme to be used commercially on a large scale (Fig. 7.3).

Initially over 40 different immobilization procedures were tried but most were discarded for one or more reasons, e.g. instability of the support matrix, low yield and/or low activity of the preparation. The three most promising methods which emerged were ionic binding to DEAE-Sephadex, covalent bonding to iodoacetyl cellulose, and entrapment within polyacrylamide. A detailed comparison of the operational properties of the enzyme immobilized by those three methods is shown in Table 7.1. It is worth noting that some of the kinetic properties of the immobilized enzyme preparations are quite different from those of the free enzyme; this is a common observation and frequently there is no obvious explanation. From the data in Table 7.1 it can be seen that there is little from which to choose between the performances of the various preparations, with the possible exception of the long half-life of the aminoacylase immobilized on DEAE-Sephadex. However, once cost factors and ease of preparation

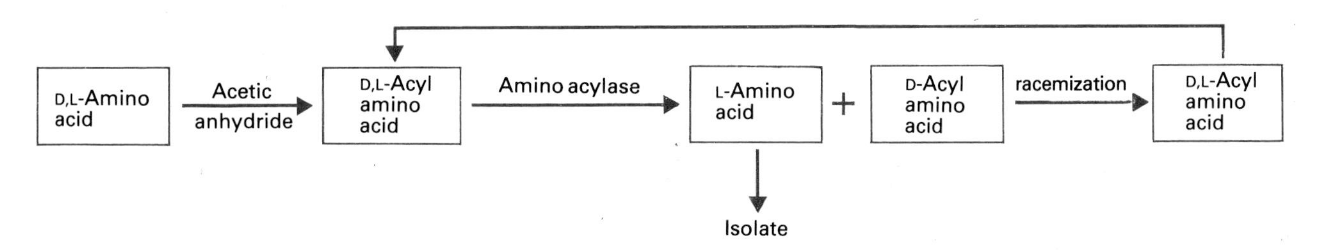

Fig. 7.2 Resolution of amino acid racemates by acylation and selective (enzymic) deacylation.

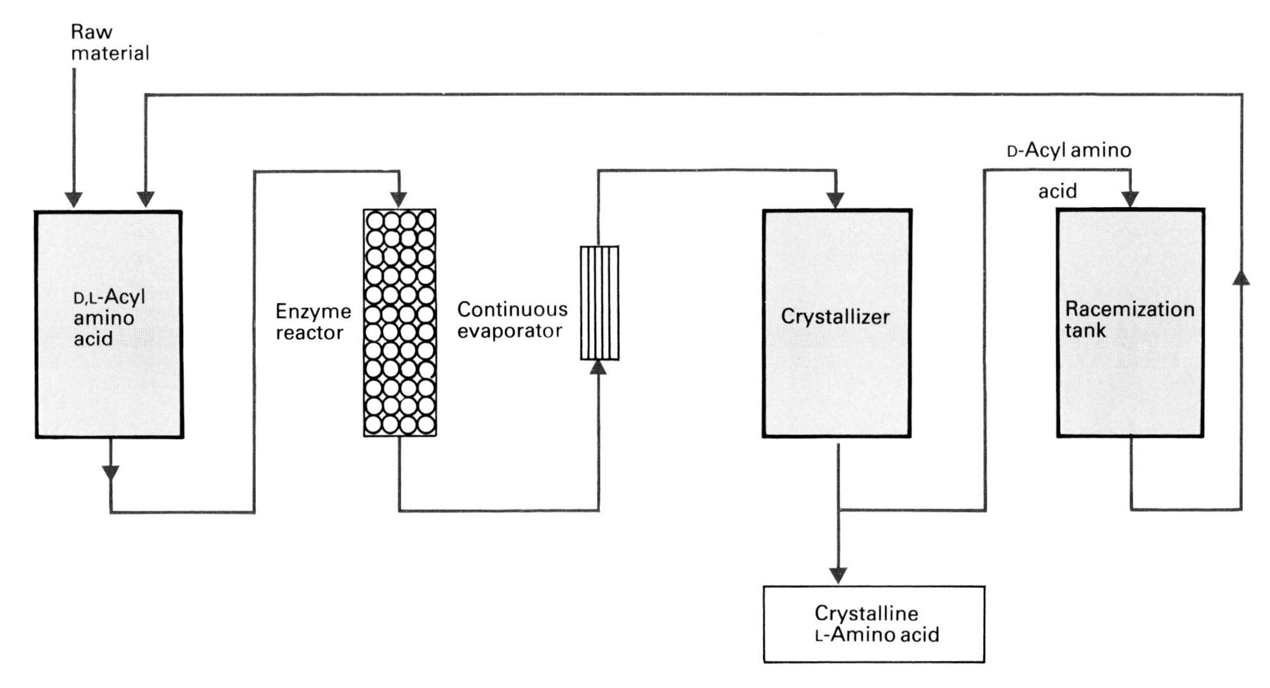

Fig. 7.3 Schematic representation of the commercial process for amino acid resolution.

Table 7.1 Comparison of properties of free and immobilized aminoacylase.

Physicochemical property	Free aminoacylase	Method of immobilizing aminoacylase: DEAE-Sephadex	Iodoacetyl cellulose	Polyacrylamide entrapment
pH optimum	7.5–8.0	7.0	7.5–8.0	7.0
Temperature optimum	60 °C	72 °C	55 °C	65 °C
Activation energy at 37 °C and pH 7.0	6.9 Kcal/mol	7.0 Kcal/mol	3.9 Kcal/mol	5.3 Kcal/mol
K_m at 37 °C and pH 7.0	5.7 mmol/l	8.7 mmol/l	6.7 mmol/l	5.0 mmol/l
V_{max} at 37 °C and pH 7.0	1.52 mol/h	3.33 mol/h	4.65 mol/h	2.33 mol/h
% Retention of activity after 10 min at 70 °C	12.5	87.5	62.5	34.5
Half-life	—	65 days at 50°C	—	48 days at 37 °C

are included in the assessment (Table 7.2) selection becomes easier. A key feature of the use of DEAE-Sephadex for immobilization is that it can be recycled when the activity of the preparation falls below the desired level, and this led to the choice of the DEAE-Sephadex immobilized aminoacylase for commercial development. In practice, the activity of an enzyme column drops by 40% during 30 days operation at 50 °C but is regenerated by the addition of fresh enzyme.

A detailed analysis of the recurrent costs associated with the immobilized aminoacylase system has shown them to be more than 40% less than those incurred with the original soluble enzyme process (Fig. 7.4). As would be expected, considerable savings can be made because the enzyme is used for long periods and is not discarded and, because the immobilized enzyme process is continuous, it can be automated and this, in turn, reduces labour costs. The savings in substrate cost are due to the ease of recovery of product which results in higher yields.

Table 7.2 Factors to be taken into account in determining method of choice for immobilization of aminoacylase.

Factor	Method of immobilization of aminoacylase		
	DEAE-Sephadex	Iodoacetyl cellulose	Polyacrylamide entrapment
Ease of preparation	Easy	Difficult	Intermediate
Retention of activity on immobilization	High	High	High
Cost of immobilization	Low	High	Moderate
Strength of binding	Medium	Strong	Strong
Operational stability (half-life)	Long	—	Moderate
Ability to be regenerated	Possible	Not possible	Not possible

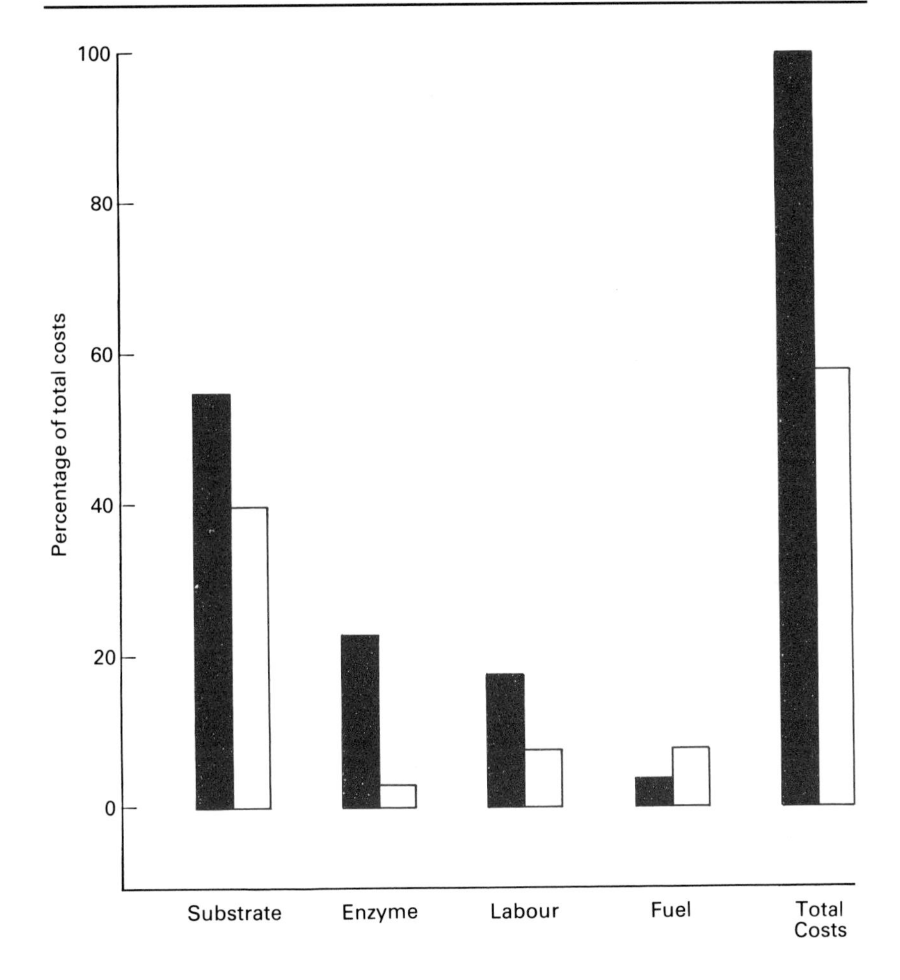

Fig. 7.4 Comparison of costs associated with soluble (solid red bars) and immobilized (open red bars) enzyme processes for aminoacylase resolution of D,L-amino acids.

OTHER APPLICATIONS OF IMMOBILIZED CATALYSTS

Apart from the resolution of D,L-amino acids, immobilized catalysts are used for the commercial production of a number of other chemicals (Table 7.3). Among these is the best example of the commercial realization of immobilized biocatalyst technology, the production of high-fructose corn syrup (HFCS). The production of HFCS involves the use of two amylases and glucose isomerase. The amy-

Table 7.3 Commercial application of immobilized cells and enzymes.

Product	Use	Enzyme and reaction	Catalyst
L-Amino acids	Various	RCHCOOH (NH-COR′) D,L-acyl amino acid —aminoacylase→ R-CHCOOH (NH_2) L-amino acid + RCHCOOH (NH-COR′) D-acyl amino acid	Enzyme immobilized on DEAE-Sephadex
High-fructose syrup	Fructose approx. 1.5 times sweeter than sucrose	100% glucose syrup (from starch) —glucose isomerase→ 42% fructose, 50% glucose, 8% other sugars	Enzyme immobilized on cellulose ion-exchanger
6-Amino penicil-lanic acid	Chemical modification to produce semi-synthetic penicillins	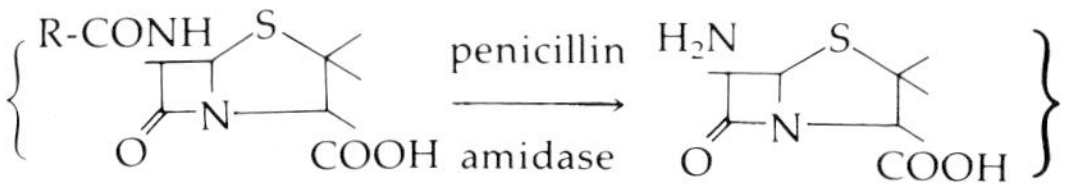 R-CONH–(S, N, O, COOH ring structure) naturally produced penicillins —penicillin amidase→ H_2N–(S, N, O, COOH ring structure) 6-amino penicillanic acid	Enzyme immobilized on sepharose with cyanogen bromide or *E. coli* cells entrapped in polyacrylamide gels
Urocanic acid	Sunscreen for use in suntan lotion	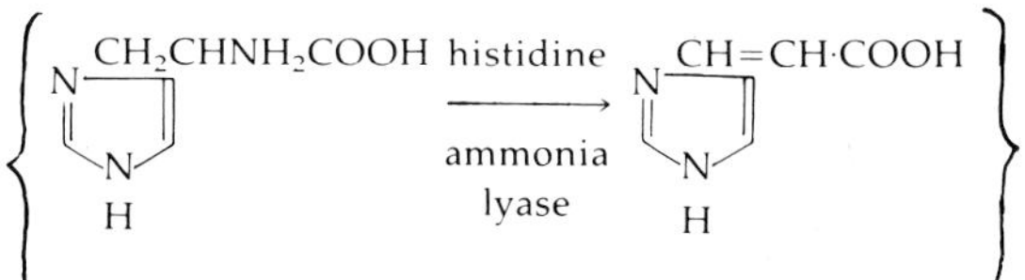 CH_2CHNH_2COOH (imidazole ring, N, N–H) histidine —histidine ammonia lyase→ CH=CH·COOH (imidazole ring, N, N–H) urocanic acid	*Achromobacter liquidum* immobilized in polyacrylamide
Malic acid	Acidulant in food industry	COOH–CH=CH–COOH fumaric acid —fumarase→ COOH–CHOH–CH_2–COOH malic acid	*Brevibacterium flavium* immobilized in carrageenan
Aspartic acid	Production of the artificial sweetener aspartame	COOH–CH=CH–COOH fumaric acid —aspartase, + NH_3→ COOH–$CHNH_2$–CH_2–COOH aspartic acid	*E. coli* cells immobilized in carrageenan

lases effect the liquefaction and subsequent degradation (*saccharification*, meaning sugar formation) of the cornstarch to glucose. The glucose isomerase then converts the glucose to fructose until equilibrium is reached when the two are approximately equimolar. The interest in HFCS stems from the fact that fructose is sweeter than glucose yet has the same calorific value; the inclusion of HFCS instead of glucose syrup in processed foods permits the same level of sweetness but provides fewer calories. In the period 1970–80 HFCS increased its share of US per capita consumption of nutritive (i.e. calorie-generating) sweeteners from almost nil to over 16%. In the same period sucrose usage decreased from

84% to 68%. By 1981 over 2.5 million metric tons (dry weight) of HFCS were produced.

The increasing demand for low-calorie sweeteners is responsible for the rapid development of another immobilized biocatalyst, that involved in the production of L-aspartic acid. The interest in aspartic acid stems from the fact that it is an ingredient of aspartame (aspartyl phenylalanine methyl ester). Since it is 160 times sweeter than sucrose only trace amounts of aspartame need to be added to food and thus it is considered to be non-nutritive, i.e. it contributes little to the calorific content of the food. In 1981 the US Food and Drug Administration approved the use of aspartame as a food additive. Since then sales of aspartame, and hence aspartic acid, have grown rapidly to thousands of tonnes per year. Indeed, the spectacular increase in the use of aspartame may eventually lead to a decline in the use of HFCS!

Another important chemical produced with the aid of an immobilized biocatalyst is 6-aminopenicillanic acid. This is a key intermediate in the production of the semi-synthetic penicillins (Fig. 7.5), which are the mainstay of current antimicrobial chemotherapy. The original fermentations of *Penicillium chrysogenum* (see p. 6) produced a mixture of penicillins but by appropriate medium supplementation could be induced to synthesize benzylpenicillin almost exclusively. Benzylpenicillin revolutionized the treatment of many potentially lethal bacterial infections but in spite of this was a less than perfect antimicrobial agent. In particular, it exhibited a restricted antibacterial spectrum, it had to be administered by injection since it was broken down by gastric acidity when given orally, and it was hydrolysed by enzymes known as β-lactamases which are produced by many bacteria.

A quite different application of immobilization,

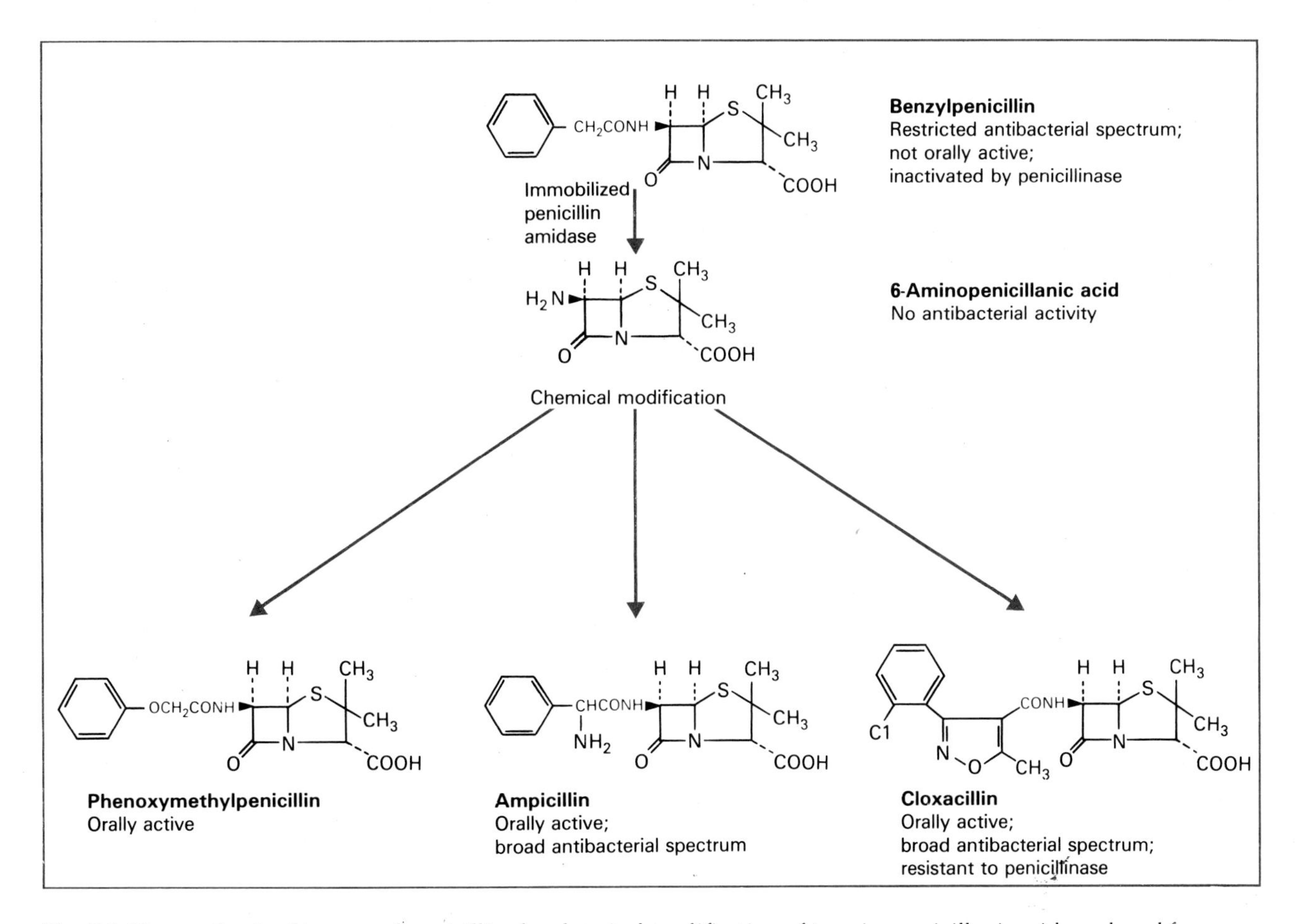

Fig. 7.5 The synthesis of improved penicillins by chemical modification of 6-aminopenicillanic acid produced from benzylpenicillin by immobilized *E. coli* cells containing penicillin amidase.

which is still in its infancy, is the manufacture of enzyme electrodes. These are probes which are capable of generating an electrical potential as a result of a reaction catalysed by an immobilized enzyme that is fixed onto or around the probe. The general principle of enzyme electrodes is embodied in the glucose electrode. In it, glucose oxidase is immobilized in a polyacrylamide gel which surrounds an oxygen electrode. The operating principle is that the enzyme catalyses the removal of oxygen from solution at a rate proportional to the concentration of glucose present. A whole series of enzyme electrodes have been constructed but to be of practical value they must be easy to make, cheap, stable, sensitive and have a low response time. Electrodes which fulfil these criteria are available for many of the commercially important microbial metabolites, e.g. penicillin and various amino acids, and, in theory, could be invaluable for on-line analysis of fermentation broths (see p. 55). In practice, they are neither robust enough nor can they be sterilized.

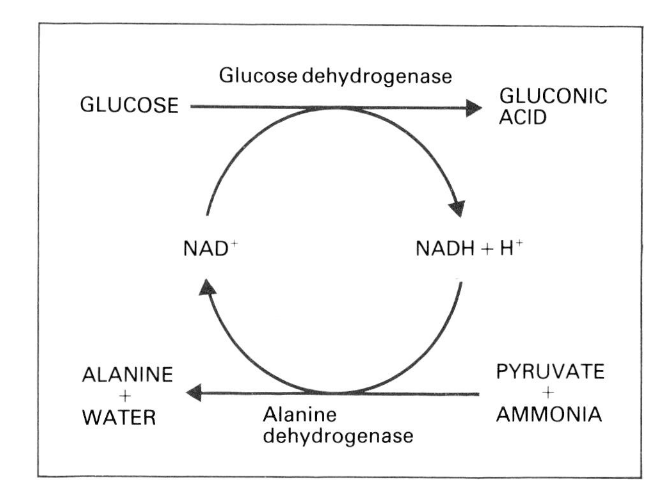

Fig. 7.6 A coupled enzyme reaction for the recyling of a pyridine nucleotide co-factor.

THE CHOICE BETWEEN IMMOBILIZED CELLS AND IMMOBILIZED ENZYMES

In some instances there is no difficulty in choosing between cells or enzymes for immobilization. For example, if the enzyme with the desired activity is secreted from the cell, then clearly it is the enzyme which should be immobilized. With protein engineering it is even possible to combine several enzymic activities in a single protein molecule. If activity of the enzyme is dependent on its spatial arrangement in the cell membrane, then clearly it is the cell which should be immobilized. More often there is little to choose between the two systems and then the overriding factor is cost. In this context it is not the cost of immobilization but the sum of the capital and recurrent costs associated with initiating and running the process. Of particular importance are the retention of activity on immobilization and the operational half-life.

To date the commercially developed immobilized systems (Table 7.3) are ones in which the catalyst has no requirement for a co-factor. The commonest co-factors are $NAD^+/NADP^+$ and ATP and their cost is such that it is not economically feasible to add them continuously with the substrate. The only practical proposition is to recycle the co-factors. This can be done in enzyme reactors (Fig. 7.6) but usually two substrates must be added and two products are obtained, only one of which is required. In addition, the co-factor also must be immobilized or it will be lost from the system and such immobilized co-factors tend to have a short half-life. Immobilized cells may be more useful in this respect for, if the cell membrane remains intact during the immobilization step, the co-enzymes are retained as are all the enzymes necessary for recycling them. However, if the cell membrane does remain intact then diffusion of substrate and product across it may be rate-limiting.

SOME OPERATIONAL CONSIDERATIONS

A number of factors can affect the operational stability of immobilized cell and enzyme reactors; one of these is the purity of the substrate used. Impurities are important for two reasons. First, they can accumulate during prolonged reactor operation and either foul the reactor bed or, worse still, irreversibly inactivate the catalyst. Second, these impurities can act as substrates for growth of microbial contaminants. Growth of contaminants can be minimized by the use, wherever possible, of high temperatures and pH values and high substrate concentrations.

A different problem is gas exchange. In some immobilized cell systems the viability of the cells has to be maintained and this can mean providing a supply of oxygen. This can best be achieved by using a fluidized bed. In such a system the beads of immobilized cells are kept in suspension by the introduction of air into the bottom of the reactor. Clearly a design has to be selected in which the immobilized cells can be prevented from leaving the reactor in the effluent stream. An alternative situation is one in which a gas such as carbon dioxide is a product of the reaction. Here the gas can be stripped out of the reactor by using high substrate feed rates (*medium velocity*).

A factor which can preclude the use of immobilized biocatalysts is the generation of heat. It is not easy to design reactors for immobilized cells and enzymes which efficiently remove the heat produced by an exothermic reaction. Thus the heat generated in the middle of an aggregate of immobilized catalyst may be sufficient to inactivate the enzyme. Since the heat produced will be proportional to the substrate concentration, one solution is to reduce the concentration of substrate in the feed solution. However this will reduce the concentration of product in the reactor effluent making product recovery more difficult and expensive and the process may not be economic. It should be noted that the generation of heat might not be observed on the laboratory scale because the surface area to volume ratio of the reactor is such that the heat produced is dissipated by conduction and convection. Thus before embarking on construction of a large-scale reactor it is advisable to check the thermodynamics of the reaction in a calorimeter.

ROLE OF RECOMBINANT DNA TECHNOLOGY

In many instances biological processes have not been developed because of the high cost of the catalyst. Thus, enzymic hydrolysis of lactose has not been economically viable because of the costs associated with production of the enzyme β-galactosidase. In this respect recombinant DNA technology is of particular value since it is an easy matter to increase the level of an enzyme in the cell from less than 0.1% to over 20% of total cell protein. Even where a non-recombinant process is viable, increasing the enzyme level in this way should cut catalyst costs and permit greater catalyst loadings per unit volume of reactor. As indicated above, an important feature of an immobilized catalyst is its half-life. Gene manipulation can be of value in this context for, as discussed in Chapter 4, it is possible to engineer proteins such that their in-vitro stability is increased, e.g. by introducing additional disulphide bridges.

Further reading

GENERAL

Chibata I., Tosa T. & Sato T. (1979) Use of immobilized cell systems to prepare fine chemicals. In Peppler H.J. & Perlman D. (eds) *Microbial Technology*, 2nd edn, Vol. II, pp. 433-61. Academic Press, London.

Klibanov A.M. (1983) Immobilized enzymes and cells as practical catalysts. *Science* **219**, 722–7.

Laskin A.I. (ed.) (1985) *Enzymes and Immobilized Cells in Biotechnology*. Benjamin Cummings, Menlo Park.

Trevan M.D. (1980) *Immobilized Enzymes*. J. Wiley & Sons, Chichester.

SPECIFIC

Jensen S.E., Westlake D.W.S. & Wolfe S. (1984) Production of penicillins and cephalosporins in an immobilized enzyme reactor. *Applied Microbiology and Biotechnology* **20**, 155–60.

Klein M.D. & Langer R. (1986) Immobilized enzymes in clinical medicine: an emerging approach to new drug therapies. *Trends in Biotechnology* **4**, 179–86.

Kricka L.J. & Thorpe G.H.G. (1986) Immobilized enzymes in analysis. *Trends in Biotechnology* **4**, 253–8.

Vandamme E.J. (1983) Peptide antibiotic production through immobilized biocatalyst technology. *Enzyme and Microbial Technology* **5**, 403–15.

Wood L.L. & Calton G.J. (1984) A novel method of immobilization and its use in aspartic acid production. *Bio/technology* **2**, 1081–4.

Part IV
Animal Biotechnology

8/Principles of Mammalian Cell Culture

INTRODUCTION

The cultivation of animal tissue *in vitro* was first shown to be possible in 1907 but the reproducible and reliable large-scale culture of mammalian cells has been achieved only in the last 10–15 years. Originally large-scale culture was developed for the production of animal viruses to be used as vaccines. More recently the goal has been the production of human proteins of potential therapeutic value and monoclonal antibodies; these topics are discussed in the next two chapters.

ESTABLISHING CELLS IN CULTURE

The cultivation of mammalian cells from a particular tissue begins with the dissociation of the tissue fragment into its component cells by treatment with a proteolytic enzyme, usually trypsin. After removal of the trypsin the cell suspension is placed in a flat-bottomed glass or plastic container together with a suitable liquid medium. After a lag period the cells attach themselves to the bottom of the container and start dividing mitotically. A culture of this type, arising directly from differentiated tissue, is referred to as a *primary culture*. Eventually the bottom of the culture vessel will be covered with a continuous layer of cells, often one cell thick and hence referred to as a *monolayer*.

Primary cultures are usually prepared from relatively large tissue masses and so will comprise a variety of differentiated cells, e.g. fibroblasts, macrophages, epithelial cells, lymphocytes, etc. However, the cells that multiply best in culture have the spindle shape (Fig. 8.1(a)) and growth rate of connective tissue cells and are called *fibroblasts*. Since epithelial cells possess the enzyme D-amino acid oxidase whereas fibroblasts do not, it is possible selectively to promote the growth of epithelial cells in culture by using D-valine as the sole source of valine. Such epithelial cells are polygonal in shape (Fig. 8.1(b)). However, the shape of cells in culture is not invariable and can be affected by the composition of the medium and the presence of infectious agents.

The cells of primary cultures can be detached from the culture vessel by trypsin treatment or the addition of the chelating agent EDTA. These cells can be used to initiate *secondary cultures* by reseeding them in fresh media at high density. Cells from primary cultures can often be transferred serially a number of times. The cells multiply at a constant rate over successive transfers and such cells comprise a *cell strain*. Cell strains do not have infinite life and divide only a finite number of times before their growth rate declines and they die; for example, human cells generally divide only 50–100 times before dying.

CONTINUOUS CELL LINES

Not all animals yield primary cultures all of whose secondary progeny die. Most murine cells die after dividing 30–50 times but a few cells become altered such that they acquire a different morphology, grow faster and are able to start a culture from a smaller number of cells. Progeny derived from such exceptional cells have unlimited life, unlike the cell strains from which they originated, and are designated *cell lines*.

During repeated serial transfer cell lines can undergo extensive changes in their cultural properties; for example, the density at which division ceases may increase such that the cells grow in clumps rather than in monolayers and the cells may be irregularly oriented in respect to each other. Such cell lines are said to be *transformed* and generally are *neoplastic*, that is they produce cancer if transplanted into related animals. Lines of transformed cells can also be obtained from normal primary cell cultures or cell strains by infecting them with oncogenic viruses or treating them with carcinogenic chemicals. Whereas no human cell line has been shown unambiguously to have originated from normal cells a number have been derived from

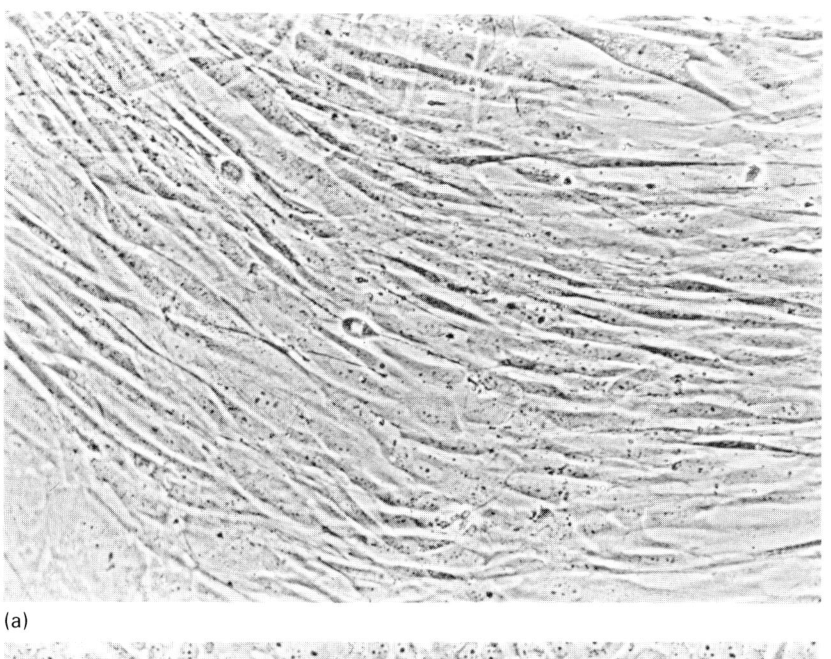

(a)

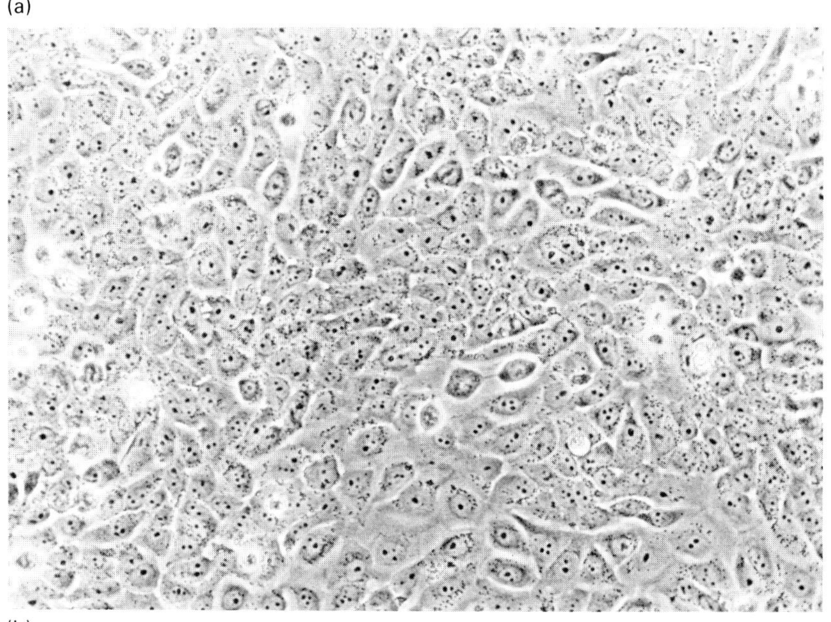

(b)

Fig. 8.1 Morphology of human cells in culture. (a) Fibroblasts; (b) epithelial cells.

human tumours. Apparently the possession of the cancerous phenotype allows easier adaptation to growth in cell culture. This may be related to the fact that cancer cells, unlike those of cell lines derived from primary cultures, generally possess extra chromosomes, that is, they are *aneuploid*. The properties of a number of the common cell lines are shown in Table 8.1.

The tendency of cell lines to change continuously on repeated cultivation necessitates that stocks of cells generally be maintained in the frozen state. The cells are mixed with additives such as glycerol or dimethyl sulphoxide to minimize cellular damage by ice crystals, are dispensed in ampoules and stored in liquid nitrogen. Cells maintained in this way remain viable for years and on thawing readily initiate new cultures.

Just as the cultural properties of cell lines may be different from those of normal cells, so too is their metabolism; for example, cells from liver-derived hepatomas fail to synthesize most of the normal liver enzymes but may continue to secrete the major liver product albumin. Similarly, cell lines from some tumours secrete hormones in an uncontrolled fashion,

Table 8.1 Properties of some commonly used mammalian cell lines.

Cell line	Species of origin	Tissue of origin	Morphology	Ploidy	Growth in suspension
3T3	Mouse	Connective tissue	Fibroblast	Aneuploid	No
L	Mouse	Connective tissue	Fibroblast?	Aneuploid	Yes
CHO	Chinese hamster	Ovary	Epithelial	Quasidiploid	Yes
BHK21	Syrian hamster	Kidney	Fibroblast	Diploid	Yes
HeLa	Human	Cervical carcinoma	Epithelial	Aneuploid	Yes
WISH	Human	Amnion	Epithelial	Aneuploid	?
Hep-2	Human	Carcinoma of larynx	Epithelial	Aneuploid	?
KB	Human	Nasopharyngeal carcinoma	Epithelial	Aneuploid	Yes

e.g. cells from tumours of the testes secrete androgen while pituitary tumour cell lines secrete adrenocorticotrophic hormone and growth hormones.

MEDIA FOR MAMMALIAN CELL CULTURE

One of the most crucial factors in achieving the successful cultivation of mammalian cells *in vitro* is the composition of the growth medium. To be satisfactory the medium must fulfil the criteria set out in Table 8.2.

The growth media in common use are complex and generally not completely defined because of the presence of serum, although this can sometimes be replaced simply with albumin. The basic component of these media is a balanced salt solution whose functions are to provide:

1 essential inorganic ions;
2 the correct osmolality;
3 the correct pH;
4 a source of energy in the form of glucose;
5 a pH indicator, phenol red.

The compositions of two widely used balanced salt solutions are shown in Table 8.3.

Although cells will remain alive for several hours in a balanced salt solution, many more ingredients must be included if proliferation of cells is to occur. These ingredients include most of the common amino acids, eight vitamins (Table 8.4) and 5–10% of serum. The precise contribution made by the serum is not fully understood. Some components of

Table 8.2 Criteria for a successful culture medium for animal cells.

1 The medium must provide all nutritional requirements of the cell
2 The medium must maintain a pH value of 7.0–7.3 despite the production of acid, i.e. it must be adequately buffered
3 The medium must be isotonic with the cell cytoplasm
4 The medium must be sterile

Table 8.3 Comparison of two balanced salt solutions.

	Amount (mg/l) in balanced salt solution of	
Ingredient	Earle	Hanks
NaCl	6800	8000
KCl	400	400
$CaCl_2.2H_2O$	264	185
$MgSO_4.7H_2O$	200	200
$NaH_2PO_4.H_2O$	140	—
Na_2HPO_4	—	47.5
KH_2PO_4	—	60
$NaHCO_3$	1680	350
Glucose	1000	1000
Phenol red	17	17

Table 8.4 Defined supplements added to balanced salt solutions for growth of HeLa cells. Figures in parentheses are for optimal growth of mouse fibroblasts.

Ingredient	Amount (mmol/l)
L-arginine HCl	0.1
L-cysteine.2HCl	0.05 (0.02)
L-glutamine	2.0 (1.0)
L-histidine HCl.H_2O	0.05 (0.02)
L-isoleucine	0.2
L-leucine	0.2 (0.1)
L-lysine HCl	0.2 (0.1)
L-methionine	0.05
L-phenylalanine	0.1 (0.05)
L-threonine	0.2 (0.1)
L-tryptophan	0.02 (0.01)
L-tyrosine	0.1
L-valine	0.2 (0.1)
D-calcium pantothenate	10^{-3}
Choline chloride	10^{-3}
Folic acid	10^{-3}
i-Inositol	10^{-3}
Nicotinamide	10^{-3}
Pyridoxal.HCl	10^{-3}
Riboflavin	10^{-4}
Thiamin.HCl	10^{-3}

the serum, e.g. α-globulins and hormones, may exert a beneficial effect by promoting the attachment and spreading of the cells and by stimulating cell division. The most popular source of serum is from fetal calves since this maintains better growth of cultured cells than serum from adult animals. Not surprisingly, fetal calf serum is prohibitively expensive and generally in short supply. Serum must be obtained from a reputable supplier who has a rigorous quality control programme (Table 8.5). A few cell lines can be maintained in media devoid of any natural components but these must be considered the exception rather than the rule. More success has been achieved when the serum has been replaced with albumin and supplements such as insulin, selenium and transferrin.

Buffering of culture media for mammalian cells is usually provided by sodium bicarbonate. This dissociates in solution releasing carbon dioxide into the atmosphere and hydroxyl ions into the medium. If dissociation occurs faster than acid is produced by the cells, then the medium will turn alkaline. Thus the buffering capacity of the culture medium is better maintained if the proportion of the CO_2 in

Table 8.5 Some of the quality control tests carried out on high-quality serum for use in mammalian cell culture.

Test	Desired result
Sterility	No growth on a variety of media (e.g. blood agar, thioglycollate broth, Sabourand's dextrose slant) at room temperature or 37 °C
Mycoplasmas and viruses	Absent
Efficacy of cell growth	Compare cell growth in T-flask with control serum lot
Total protein content and immunoglobulin electrophoresis pattern	Same as content/pattern of control serum
pH value after filtration	7.2–7.8
Osmolality	290–310 mosm/l
Endotoxin content	Less than 2 ng/ml by the Limulus Amoebocyte Lysate test
Haemoglobin content	Less than 20 mg%

the gas phase inside the culture vessel is high. This is reflected in the differences in bicarbonate content of the two balanced salt solutions shown in Table 8.3. Cells which release a lot of metabolic CO_2 are best cultured in supplemented Hanks' saline in a sealed container where the CO_2 which they release helps slow down the dissociation of the sodium bicarbonate. Cells which do not produce much CO_2 do better in Earle's saline exposed to a gas mixture of 5% CO_2 in air. Because of the difficulties associated with the use of bicarbonate buffers a number of organic buffers have been introduced for animal cell culture. The most popular of these is HEPES (N-2-hydroxyethyl piperazine-N'-2-ethanesulphonic acid) which has a pKa of 7.3 at 37°C. Unfortunately this compound is too expensive for routine large-scale use.

The media used to culture animal cells are rich in nutrients so they are particularly susceptible to contamination with bacteria and fungi unless good aseptic technique is practised. For this reason antibiotics are usually added: penicillin to prevent growth of Gram-negative cells and nystatin to inhibit fungi. Another problem with cell cultures is contamination with mycoplasmas, a group of bacteria which lack cell walls. Most of the contaminating mycoplasmas are non-pathogenic and probably originate from the oropharynx of laboratory staff, but some are derived from trypsin or serum. Tetracyclines, kanamycin and gentamycin can be used to inhibit most mycoplasmas.

PREPARATION OF MEDIA AND EQUIPMENT FOR CELL CULTURE

Probably the most important aspect of animal cell culture is the quality of the water used. High-quality distilled water is required, and this is usually prepared from deionized water. The water used for washing and rinsing of equipment must be of the same quality, as should be the steam used for sterilization. Common laboratory or plant steam cannot be used since it contains additives toxic to mammalian cells and is contaminated with metal ions unless stainless steel is used throughout.

All the materials which come into contact directly or indirectly with cultured mammalian cells must be chemically clean as well as sterile. Immersion in nitric or hydrochloric acid is an effective method of cleaning glassware; most detergents cannot be used since they tend to persist on glass surfaces. However cleaned, all apparatus must be exhaustively rinsed with high-quality distilled water. For sterilization of apparatus dry heat can be used but large assemblies are usually steam sterilized.

Growth medium is prepared in vessels constructed from glass or high-grade stainless steel. High-quality water of the desired volume is mixed with the chemical ingredients of the medium, serum is added and the pH value adjusted if necessary. Because some of the components of the medium will be heat labile sterilization is done by filtration.

MODES OF CELL GROWTH

Under optimal conditions primary epithelial cells form layers one cell thick while the lawns formed by fibroblasts are usually 2–3 cells thick. Such cells generally only grow when attached to surfaces. The same principles apply for cell strains and cell lines derived from normal cells. Only rarely can cell lines be derived that multiply well in suspension and these usually arise from transformed cell lines. Cultures of cells which grow well in suspension produce many more cells than those whose growth is restricted to surfaces. Transformed cells and cells of some lines can grow and form colonies on semi-solid media, e.g. agarose, and this is particularly useful when cloning cells, as in hybridoma selection (see p. 118). These different modes of growth are reflected in the methods used to culture cells in the laboratories.

SMALL-SCALE CELL CULTURE

Primary cell cultures and cell strains are grown either in plastic Petri dishes (tissue culture grade) or in plastic T-flasks (Fig. 8.2). The latter are a modification of the Carrel flask introduced by Alexis Carrel in 1923. These T-flasks are available in a

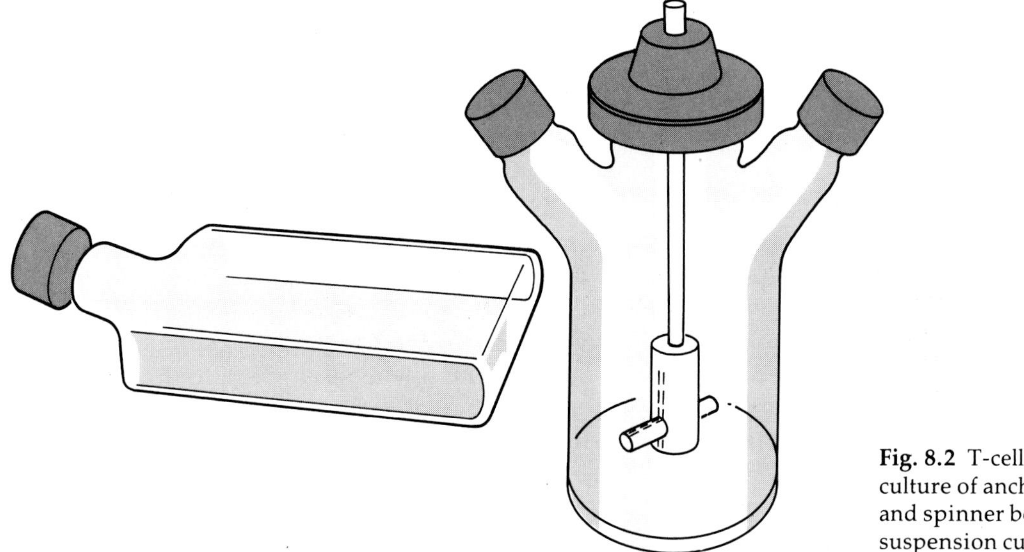

Fig. 8.2 T-cell flask (left) used for the culture of anchorage-dependent cells and spinner bottle (right) used for suspension culture.

variety of sizes with surface areas ranging from 25 to 175 cm^2. In a typical experiment a 75 cm^2 T-flask containing 15–20 ml of media would be seeded with 10^6 cells and incubated at 37 °C. After 2–3 days the bottom surface of the flask will be covered with a confluent monolayer of cells. The cells can readily be detached by trypsin or EDTA treatment.

The reason why a high cell density is required in the inoculum is related to the fact that for recently isolated cells the medium must contain adequate concentrations of unknown factors produced by the cells themselves. This is not so true for transformed cells, which can withstand far greater dilution. The efficiency of plating of recently isolated cells, e.g. primary cultures and cell strains, can be greatly increased if they are mixed with a feeder layer of similar cells made incapable of multiplication by X-irradiation; these cells are still metabolically active and supply factors that enable the non-irradiated cells to survive and multiply. The efficiency of plating can also be increased by Earle's technique of introducing individual cells into very small volumes of media, e.g. in microtitre dishes. The cell-produced factors thus reach an adequate concentration without the need for a feeder layer.

Although transformed cells can also be grown in T-flasks they can be grown more efficiently and effectively in suspension in spinner bottles. These spinner bottles (Fig. 8.2), which are simply bottles containing a captive magnetic stirrer bar, are available in a variety of sizes from 25 ml to 9 l. Bottles are inoculated with cells at a density of at least 0.25×10^6 cells/ml and purged with an air/CO_2 mixture. The contents are stirred magnetically but sufficiently gently so that shear damage to the cells does not occur; a satisfactory speed is achieved when a liquid vortex is just beginning to appear on the stirrer shaft.

LARGE-SCALE CELL CULTURE

The large-scale production of mammalian cells which grow in suspension culture is not very difficult. Standard laboratory or production fermenters can be used since these are designed to run aseptically for long periods and have facilities for gassing and pH control. However, the stirrer motor may have to be geared down to ensure that a sufficiently low speed of stirring can be obtained to prevent undesirable shear effects on the cells. An alternative method of culturing the cells is to use an airlift fermenter (p. 58); the sparged gas agitates the cells sufficiently to promote metabolite interchange without causing damage to the cells.

The growth of anchorage-dependent cells on a large scale is much more difficult, for a fermenter does not provide the requisite high surface area to volume ratio. A number of different methods have been developed and the choice of method depends partly on the use to which the cells will be put and partly on the capital which can be spent. The simplest system is the use of roller bottles which have a capacity of 500 ml to 50 l. These are filled with about one-twentieth of their volume of medium, inoculated with cells and placed on their sides on slowly revolving rollers. As the bottles revolve the growing monolayer is alternately immersed in the liquid medium and exposed to the air in the bottle. Thus the cells are automatically washed with nutrients and exposed to oxygen and the movement of the liquid medium ensures uniform distribution of its components. As can be readily envisaged, large-scale roller bottle operation involves simple technology but is labour intensive. In addition, the bottles are constructed from high-quality glass and thus are expensive but they only have a finite life; eventually cells fail to adhere to the inner surface.

One way of increasing the surface area to volume ratio in a fermenter is to fill it with a series of closely spaced plates. Various designs have been described but they all suffer from the same disadvantages: high construction costs, poor mixing and difficulty in encouraging cells to stick to the plate surfaces. A much better way of utilizing conventional fermenters is to make use of microcarrier beads. First described in 1967 by Anton van Wezel many variations of the microcarrier method have been exploited on an industrial scale. The beads are composed of dextran or a synthetic polymer and are covered with a large number of functional groups, e.g. DEAE groups. Since the beads are only 50–200 μm in diameter and have a density slightly greater than that of water they are readily maintained in suspension by gentle agitation. Anchorage-dependent cells inoculated into the medium bind to the beads and multiply. The advantage of this system over conventional roller bottles can be seen by reference to Table 8.6.

A completely different method of growing anchorage-dependent cells involves the use of hollow-

Table 8.6 Comparison of different culture methods for anchorage-dependent cells.

Roller bottle	Surface area	4.9×10^2 cm²/bottle
Microcarriers	Surface area	6×10^3 cm²/g dry wt
	Normal density	5 g dry wt/l
	Surface area/l	3×10^4 cm²
	Roller bottle equivalents	60
Hollow-fibre apparatus	Size	$40 \times 40 \times 4.5$ cm
	Fibre surface area	9.3×10^3 cm²
	Roller bottle equivalents	19

Table 8.7 Problems associated with large-scale mammalian cell culture.

1 Cost and availability of good-quality mammalian serum
2 Potential contamination of bovine serum with foot and mouth disease virus (FMDV). This is not a problem in the US, which is FMDV free
3 Potential contamination of cells with mycoplasmas or even animal viruses of human origin. Such infections are not always easy to detect
4 Large-scale use of antibiotics can be costly and they can contaminate the final product. In the case of penicillin residues this would not be acceptable
5 Require large amounts of high-quality distilled water and steam which should be pyrogen free. This is costly
6 May require special equipment or culture methods for anchorage-dependent cells

fibre technology. The cells are grown on the outer surface of the fibres which are immersed in the growth medium while air and carbon dioxide are passed through the lumen of the fibres. In practice, flat bed structures are used with the fibres no more than three to six layers deep. This arrangement ensures that when medium is pumped *through* the bed the path of perfusion is short. Thus there is hardly any gradient from a high to a low nutrient level and from a low to a high waste product level. Once the cells have encrusted the fibres it is possible to switch from a growth medium to a maintenance medium. The latter medium contains no serum, hence it is much cheaper, but it still permits the secretion of macromolecules for prolonged periods.

PROBLEMS ASSOCIATED WITH LARGE-SCALE CELL CULTURE

From the aforegoing section it should be clear that the large-scale culture of animal cells can be achieved but that it is not without its problems. Some of these problems are summarized in Table 8.7.

GENERAL GUIDELINES FOR THE USE OF ANIMAL CELL CULTURES

Listed below are some specific recommendations for the handling of animal cell cultures and which are not covered elsewhere in this chapter.

1 The cell culture facility should be designed specifically for the purpose and should be adequately equipped to permit proper aseptic laboratory procedures.
2 Whenever possible, cell lines should be obtained only from reputable repositories who should be able to provide details of the history of the culture and the quality control checks which have been performed on it.
3 When new cell lines arrive in the laboratory they should be held in quarantine, preferably in a separate area, until sterility and other tests have been concluded. Mycoplasma assays are particularly important and should include positive and negative controls.
4 Once a new culture has been shown to be free of adventitious agents, a series of stocks should be frozen down.
5 Medium components, epecially serum, should be pretested for growth promotion with appropriate cell cultures.
6 All contaminated media and infected cultures should be disposed of promptly and properly.
7 Cell lines should be checked routinely to ensure that they exhibit no changes in karyotype, phenotypic expression or growth pattern.

Further reading

GENERAL

Feder J. & Tolbert W.R. (1983) The large-scale cultivation of mammalian cells. *Scientific American* **248**, 24–31.
Harakas N.K. (1984) Industrial mammalian cell culture: Physiology-technology-products. *Annual Report of Fermentation Processes* 7, 159–211.

Jakoby W.B. & Pastan I.H. (eds) (1979) *Methods in Enzymology*, Vol. 58, Cell Culture. Academic Press, New York.

Thilly W.G. (ed.) (1986) *Mammalian Cell Technology*. Butterworths, London.

SPECIFIC

Clark J.M. & Hirtentein M.D. (1981) Optimizing culture conditions for the production of animal cells in microcarrier culture. *Annals of the New York Academy of Sciences* **369**, 33–46.

Eagle H. (1955) Nutrition need of mammalian cell in tissue culture. *Science* **122**, 501–4.

Hopkinson J. (1985) Hollow fibre cell culture systems for economical cell-product manufacturing. *Bio/technology* **3**, 225–30.

Ku K., Kuo M.J., Delente J., Wildi B.S. & Feder J. (1981) Development of a hollow fibre system for large-scale culture of mammalian cells. *Biotechnology and Bioengineering* **23**, 79–95.

McKeehan W.L. (1986) Growth factors spawn new cell cultures. *Nature* **321**, 629–30.

9/Applications of Animal Cell Culture

INTRODUCTION

Scientists studying the biochemistry and biophysics of cell growth and division have made extensive use of cultured animal cells, as have those engaged in research on animal viruses. However, the value of cultured animal cells goes far beyond this, for they produce a wide range of biological products of commercial interest including immunoregulators, antibodies, polypeptide growth factors, enzymes and hormones. Already they are used in the manufacture of virus vaccines, tissue plasminogen activator (an enzyme which facilitates the destruction of blood clots), interferon-α (for the treatment of cancer), monoclonal antibodies and tumour-specific antigens (for inclusion in diagnostic kits). Consideration also is being given to the possibility of growing in culture fibroblasts from, for example, burns patients. These cultured cells would be used as a reconstituent skin to facilitate wound healing. Another potential application of cultured cells is their use in evaluating new drugs and toxic chemicals. When used in concert with conventional animal models they could reduce much of the time, cost and effort as well as the total number of animals.

PRODUCTION OF VIRAL VACCINES

The oldest commercial application of animal cell cultures is for the production of viral vaccines and it could be argued that globally, with the possible exception of antibiotics, they have provided greater benefit than any other pharmaceutical product. The deployment of a single vaccine, that against smallpox, has been of immense benefit to mankind and hopefully has eliminated the virus for evermore. Because viruses are obligate intracellular parasites the earliest viral vaccines, those against smallpox and rabies, were made in intact animals such as calves, sheep and rabbits. Today the only intact host used in advanced production techniques is the developing chick embryo for the manufacture of influenza and yellow fever vaccines. All other vaccines are prepared by the growth of virus in cell culture. Table 9.1 lists the major human and veterinary virus vaccines and the cell cultures used in their production.

Viral vaccines are of two types. Those which are prepared from live, avirulent (attenuated) organisms often give lifelong protection but carry with them the attendant risk of contamination with other viruses derived from the cell culture. Clearly, rigorous microbiological and other quality control procedures are essential. Vaccines prepared from killed (inactivated) viruses are less satisfactory since they are not as immunogenic as live vaccines, presumably because the virus does not multiply in the animal host, and have to be injected repeatedly according to carefully balanced immunization schedules. The only inactivated vaccines for human use are those against rabies and, in some instances, polio. The usual inactivating agents are dilute formalin or β-propiolactone.

In theory, the production of viral vaccines is relatively simple. The cell cultures principally used for virus vaccine production are prepared from monkey kidney or chick embryos. More recently there has been an increasing use of human diploid cells. Since these cells grow as monolayers all that is required in vaccine preparation is to infect them with the appropriate virus and then harvest the culture fluid after virus multiplication has occurred. The culture fluid containing the virus is clarified by filtration. In the case of killed-virus vaccines a concentration step is usually employed after the inactivation step. When processing is complete the virus suspension is blended with stabilizers to prevent loss of potency, a problem with live attenuated vaccines, and stored at low temperature. A typical production process is shown in Fig. 9.1.

In practice, the production of a virus vaccine is far more complex, not just because of the labour involved in large-scale animal culture but because of the rigorous quality control and safety procedures which are required; for example, all tissue culture

Table 9.1 Some viral vaccines currently available for human and veterinary use.

Human vaccines		Veterinary vaccines	
Virus	Cell used for culture	Virus	Cell used for culture
Measles	Chick embryo fibroblasts	Canine distemper	Chick embryo fibroblasts or dog kidney cells
Polio (inactivated)	Monkey kidney cells	Canine hepatitis	Dog kidney cells
Polio (live)	Monkey kidney or human diploid cells	Foot and mouth disease	Bovine kidney cells
Rabies	Human diploid cells	Rabies	Duck embryo or chick embryo
Rubella	Rabbit kidney, duck embryo or human diploid cells	Feline panleucopenia	Cat kidney cells
		Marek's disease	Chick embryo cells

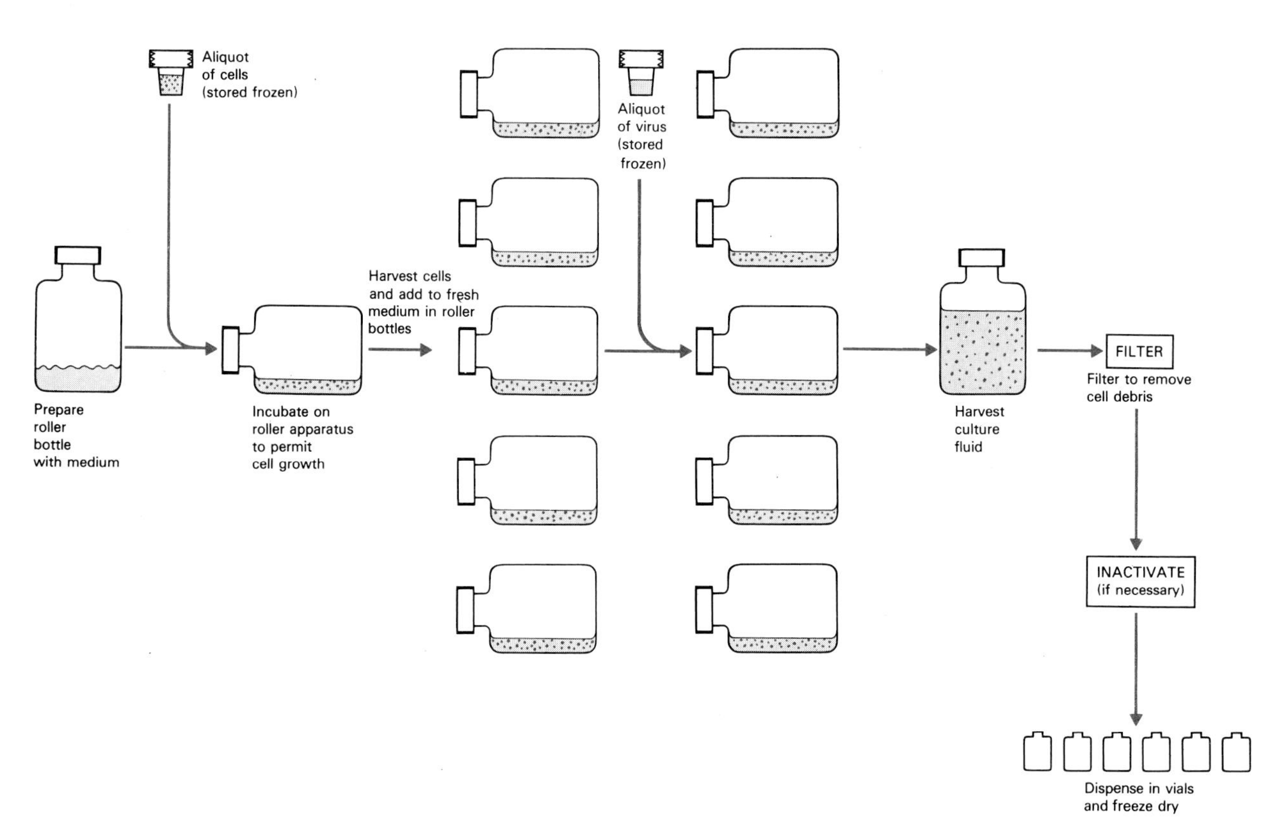

Fig. 9.1 Production of killed-virus vaccines.

substrates are rigorously examined to exclude contamination with infectious agents from the source animal or, in the case of human diploid cells, to exclude abnormal cellular characteristics. Monkey tissues are tested exhaustively for extraneous agents, e.g. Herpes virus B, Simian virus 40, tubercle bacilli and mycoplasmas. Human diploid cells are subjected to detailed karyological examination to exclude cultures with features resembling transformed cell lines or malignant tissues.

Safety of the final product is another important consideration. With killed-virus vaccines the potential hazards are those due to incomplete virus inactivation — a possible cause of recent outbreaks of foot and mouth disease in Europe! The tests used to detect live virus consist of the inoculation of susceptible tissue culture and of susceptible animals. The cultures are examined for cytopathic effects and animals for symptoms of disease and histological evidence of infection at autopsy. With attenuated viral vaccines the potential hazards are those associated with reversion of the virus to a degree of virulence capable of causing disease in vaccinees. This possibility is minimized by the use of stable virus seed stocks but virulence checks on the final product are necessary; for example, in the production of attenuated poliomyelitis vaccine the neurovirulence of each batch following intraspinal inoculation of monkeys is compared with that of a control vaccine.

Many of the problems associated with the production of viral vaccines, particularly the safety aspects, may disappear now that recombinant DNA technology can be used to enable virus subunits to be produced in bacteria or yeasts. This has been discussed already in Chapter 6. Another advantage of recombinant DNA technology is that it may permit the production of vaccines against those viruses which cannot be grown in cell culture to a titre high enough to provide sufficient antigen for effective vaccination. An example based on hepatitis B virus was presented in Chapter 6 (see p. 81).

PRODUCTION OF HIGH-VALUE THERAPEUTICS

There are many human proteins which have long been believed or known to have therapeutic potential but which are in short supply; for example, human growth hormone can be obtained from the pituitary gland of cadavers and clotting factor VII and IX from the plasma of blood donors. The supply limitations of these substances is obvious and there are other complications, e.g. the potential for acquiring AIDS from contaminated blood products. A more satisfactory alternative would be to grow in large-scale culture cells derived from the tissue which normally synthesizes the desired protein. Some examples are given in Table 9.2. The disadvantage with this method is that, generally speaking, such cultured cells only produce low levels of the therapeutic protein: interferon-β is a good example. Using human embryonic lung fibroblasts growing in 20 l roller bottles, each containing 750 ml of medium, only 5×10^6 units (0.02 mg) of interferon-β are produced per litre of medium *after* induction of interferon synthesis with polyI:C. Until the advent of suitable monoclonal antibodies the purification of this interferon, free of serum proteins from the growth medium, was very difficult.

One solution to the problem of low yields is to produce the protein in bacteria by using recombinant DNA technology as outlined in Chapter 6. The problem with this approach is that many mammalian proteins undergo post-translational modifi-

Table 9.2 Proteins overproduced by some mammalian cell lines.

Product	Cell line name	Source
Acetylcholinesterase	BP4 1AS	Murine neuroblastoma
Interferon-α	Namalwa	Human blood
Interferon-β	Flow 7000	Human embryonic foreskin
Interleukin	EL4 CL14	Murine blood
Plasminogen activator	GPK	Guinea pig heratocyte
	BEB	Human breast
Urokinase	HT 1080	Human fibrosarcoma

cations by processes such as glycosylation which are not mimicked in bacterial systems. Although many of the non-glycosylated derivatives of these proteins retain activity, e.g. interferon-γ, urokinase and interleukin-2, their half-life *in vitro* may be reduced significantly and long-term therapy in humans may result in undesirable immunological reactions. With some proteins, such as human blood clotting factor IX, the situation is more complex: not only is glycosylation essential for activity but so too is the vitamin K-dependent α-carboxylation of glutamic acid residues. Clearly, at least for the foreseeable future, the only prospect of manufacturing such molecules is to do so in mammalian cells.

If mammalian cell culture is the only method for obtaining certain therapeutic proteins, then cells which are overproducers are required. It is well known that in some pathological conditions certain cells in the body become neoplastic and overproduce a particular protein; Table 9.2 lists some of the proteins overproduced by transformed cells. However, transformed cells that will produce many of the mammalian proteins currently in demand are not available. The solution is to clone the relevant genes in mammalian cells rather than microbial cells.

CLONING IN ANIMAL CELLS: BASIC PRINCIPLES

Both bacteriophage and plasmid vectors are available for gene manipulation in bacteria but because of ease of use the latter are favoured. However plasmids do not occur naturally in animal cells and so only virus vectors are available. As for any vector, the virus genome must be easily manipulated, it must contain convenient sites for the restriction enzymes used in cloning and the location of those sites with respect to vector control sequences must be known. This latter point is particularly important since the gene sequences required for virus replication are far more complex than those required for plasmid replication. With virus vectors there can be an additional complication: the need to package the recombinant genome in viral coat protein. This can lead to size constraints on the amount of foreign DNA which a vector can carry. For this reason considerable emphasis is being placed on the development of vectors which can replicate like plasmids and thus at no stage need to be encapsidated in virus particles.

Whatever kind of vector is used there will always be a need at some stage to introduce unpackaged DNA into animal cells in culture, a procedure known as *transfection*. The most commonly used procedure is dependent upon the fact that freshly prepared co-precipitates of DNA and calcium phosphate are phagocytosed by cells. The method is not particularly efficient since only a low proportion of cells take up the exogenous DNA; at best only 1–2% of cells are infected but the efficiency can be as low as 1 cell in 10^6. When virions are produced the efficiency of transfection is not a limiting factor as the recombinant virions subsequently can be used to guarantee efficient infection.

It is not essential to have vectors capable of replication in order stably to introduce DNA into mammalian cells. Fragments of DNA can be phagocytosed by means of the calcium phosphate co-precipitation method and in at least a few of the transfected cells will become stably integrated at random in the genome. Identification of these rare cells carrying exogenous DNA is difficult unless the DNA also carries a gene for a selectable marker. A range of selectable genes is available (Table 9.3) for ligating to test genes prior to transfection. This technique has been used to obtain expression of Hepatitis B surface antigen, Herpes Simplex type I glycoprotein D and human interferon-γ in Chinese Hamster ovary cells. However, even though multiple copies of the transfected gene are present, expression levels are low and this technique is better suited to gene therapy (see Chapter 11).

VIRUS VECTORS FOR USE WITH MAMMALIAN CELLS

A number of different virus vectors are available for cloning in mammalian cells. Each vector system has its advantages and disadvantages and, as yet, no one type of vector has emerged as clear favourite. To some extent this is a reflection of the different host ranges of the different viruses but personal preference also plays a part. The best example of this comes from work on the expression of clotting factor IX where three different vector systems have been used successfully. Only a brief description of each system is presented below and for a more detailed analysis the reader should begin by consulting Old and Primrose (1985). It should be borne

Table 9.3 Some markers whose presence on exogenous DNA facilitates the selection of transfected cells.

Selectable marker	Basis of selection
Thymidine kinase (from Herpes Simplex virus)	Permits growth of cells lacking thymidine kinase
Bacterial neomycin phosphotransferase	Confers resistance to aminoglycoside antibiotics, e.g. G418
Bacterial dihydrofolate reductase	Confers resistance to methotrexate

in mind that the development of animal virus vectors is progressing at a phenomenal rate.

SV40 vectors Simian virus 40 (SV40) was the first eukaryote virus to be harnessed as a vector simply because it was the first such virus for which a complete nucleotide sequence was available. The genome of SV40 contains very little non-essential DNA so it is necessary to insert the foreign gene in place of essential viral genes in order not to exceed the size limits imposed by the viral protein coat. Consequently the recombinant genome can be propagated only in the presence of a helper virus. In the original SV40 vectors the gene that is replaced is one which is required late in infection. The procedure for using such a vector is to perform a mixed infection between the vector carrying the foreign gene insert and a helper virus carrying a temperature-sensitive mutation in a gene required early in infection. Following infection at the non-permissive temperature virus growth can occur only by complementation between the helper virus, which provides late functions, and the recombinant, which provides early functions.

An alternative SV40 vector system is one in which the foreign gene insert replaces a gene required early in infection. The advantage of this system is that although these recombinant viruses are unable to replicate in most monkey cell lines, they can do so in a special line of cells known as COS. Since helper virus is not required this means that pure recombinant virus stock can be produced easily. However, expression of foreign genes from the early promoter is much lower than from the late promoter.

A whole series of vectors has been constructed which comprise part of the *E. coli* vector plasmid pBR322 and part of the SV40 genome and which permit the exogenous gene to be placed under the control of an SV40 promoter. These vectors are grown and manipulated using *E. coli* as the host and are then transferred back into monkey cells. A similar series of vectors based on a related virus, polyoma, is available for use with cultured mouse cells.

Adenovirus vectors The major disadvantage with SV40- and polyoma-based vectors is the small size of the genome. Since packaging constraints are strict, only small gene inserts can be made. Adenoviruses are much larger than SV40 and can accommodate much more extraneous DNA. However, the larger size means that the adenovirus genome is intrinsically more difficult to manipulate since it contains multiple sites for restriction endonucleases. Using a variety of techniques vectors are being constructed in which many of these sites have been eliminated. In one set of vectors exogenous DNA is placed under the control of a transcriptional unit which is functional early in the life-cycle of the virus. Although the exogenous gene replaces a viral gene function, no helper virus is required if the recombinants are grown in a special strain of human embryonic kidney cells.

Vaccinia Vectors based on vaccinia virus are used in a slightly different way. Exogenous genes are first cloned adjacent to a vaccinia promoter present in the middle of a vaccinia-derived thymidine kinase gene which is carried on an *E. coli* plasmid (Fig. 9.2). After manipulations are complete in *E. coli* the recombinant plasmid and wild-type vaccinia virus are co-transfected into a monkey cell line. Recombination occurs between the plasmid-borne thymidine kinase gene fragment and the viral thymidine kinase gene. Recombinant virions fail to synthesize thymidine kinase and can be selected by virtue of their resistance to 5-bromodeoxyuridine.

BPV vectors Vectors based on bovine papilloma virus (BPV) are unique in that not only are they propagated as multicopy plasmids but they cause neoplastic transformation of the host cell, i.e. the cell becomes cancerous. The earliest vectors comprise a large subgenomic transforming fragment of

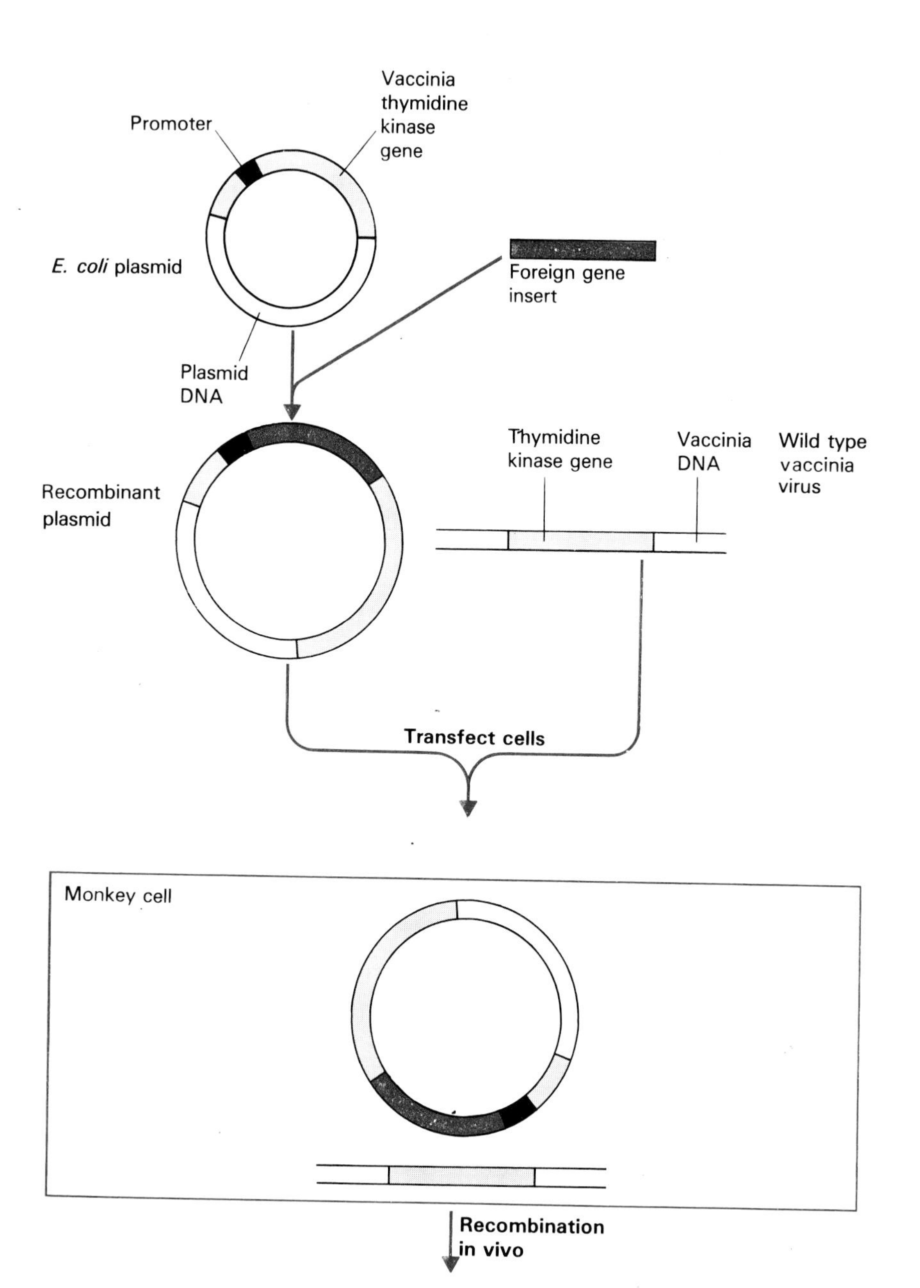

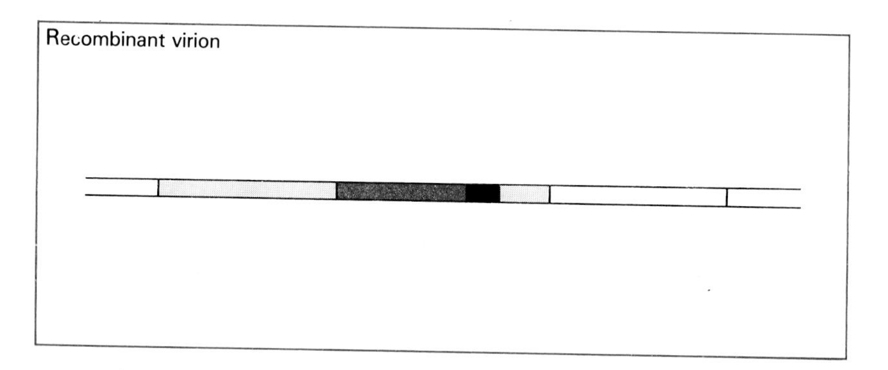

Fig. 9.2 The insertion of a foreign gene into the vaccinia genome by recombination *in vivo* (see text for details).

the BPV genome linked to an *E. coli* plasmid. Selection of cells carrying the vector is easy: being neoplastic they outgrow the normal cells which only form a monolayer in culture. BPV-derived vectors show great promise because permanent cell lines are obtained which carry the foreign DNA insert at high copy number (up to 200 per cell). An additional advantage is that there is no size limitation on the DNA insert.

Retrovirus vectors Retroviruses have single-stranded RNA genomes and, at first sight, would appear to be unpromising candidates as vectors but the peculiarities of the retrovirus life-cycle mean that these viruses provide perhaps the most promising vector system of all. During the process of reverse transcription, a double-stranded proviral DNA molecule is generated which has all the regulatory functions for viral transcription and translation (see Box, p. 112). Thus, any of the normal proviral genes can be replaced with foreign DNA provided the recombinant is propagated in the presence of a helper virus. Retroviruses have a number of features which make them particularly useful as gene transfer vectors and these are summarized in Table 9.4.

Table 9.4 Properties of retroviruses which are of value in gene cloning.

1 The normal replication process involves the stable insertion of a DNA copy of their genome into the host genome
2 They can infect and transmit their genetic information into a high proportion of recipient cells
3 Retroviruses have a broad host range which can be easily modified making it possible to infect cells from a variety of species and cell types
4 The integrated provirus exists as a stable, predictable structure within the host genome
5 Infection of mammalian cells with retroviruses does not result in cell death. Rather, infected cells continue to grow and divide while producing large numbers of virus particles
6 A large amount of foreign gene sequence can be packaged within a virus particle

PRODUCTION OF THERAPEUTICS AND VACCINES IN RECOMBINANT MAMMALIAN CELLS

A whole range of proteins of potential value to the pharmaceutical industry has now been produced in cultured, recombinant mammalian cells (Table 9.5). In some instances the levels of expression are too

Table 9.5 Some mammalian proteins overproduced in cells in culture as a result of in-vitro gene manipulation.

Protein	Size	Use	Comments
Tissue plasminogen activator	527 amino acids	Thrombosis	Enzyme overproduced in *E. coli* or yeast not fully functional. Product currently in clinical trials
Interleukin 2	133 amino acids	Cancer therapy	
Tumour necrosis factor	157 amino acids	Cancer therapy	
Factor VIII	2332 amino acids	Haemophilia	Normally obtained from plasma of blood donors but now there is concern over potential contamination with AIDS virus
Factor IX	415 amino acids	Christmas disease	Must be made in mammalian cells as glycosylation and conversion of first 12 glutamate residues to pyroglutamate essential for activity
Erythropoietin	166 amino acids	Anaemia	Without glycosylation protein is cleared very quickly from plasma

The biology of retroviruses

Retroviruses differ from all other viruses of animals in that their genome is diploid, consisting of two identical molecules of single-stranded RNA held together at their 5′-ends by hydrogen bonding. The virus particles also contain various species of low molecular weight RNA molecules including tRNA. The function of most of these smaller RNA molecules is not known but the tRNA serves as a primer for DNA replication (see Box, p. 24).

Three viral genes are essential for replication: *gag*, which encodes an internal structural protein of the virus; *pol*, which encodes an RNA-dependent DNA polymerase; and *env*, which encodes glycoproteins of the virus envelope. Viruses that carry all three of these genes are replication-competent and are termed non-defective. Many retroviruses exist which lack some or all of these essential genes and such *defective* viruses depend on non-defective helper viruses for replication. Many of the retroviruses contain a gene, *onc* or *src*, which is homologous to a normal cellular gene and whose overexpression can result in cellular transformation. Finally, most retroviruses have a *c* region which carries an efficient promoter that causes viral products to be made in large quantities. Diagram 1 shows the gene order in a typical non-defective avian retrovirus.

The replication cycle of retroviruses can be divided into three parts (Diagram 2): first, the conversion of single-stranded viral RNA to double-stranded (proviral) DNA; second, integration into the chromosome; and third, transcription and translation of the integrated provirus. In addition to being translated, the RNA which is transcribed may be packaged into virus particles.

The RNA-dependent DNA polymerase, or reverse transcriptase, encoded by the *pol* gene has two enzymatic activities: a DNA polymerase capable of copying RNA or DNA templates and a ribonuclease (RNaseH) that specifically removes the RNA from RNA: DNA hybrids.

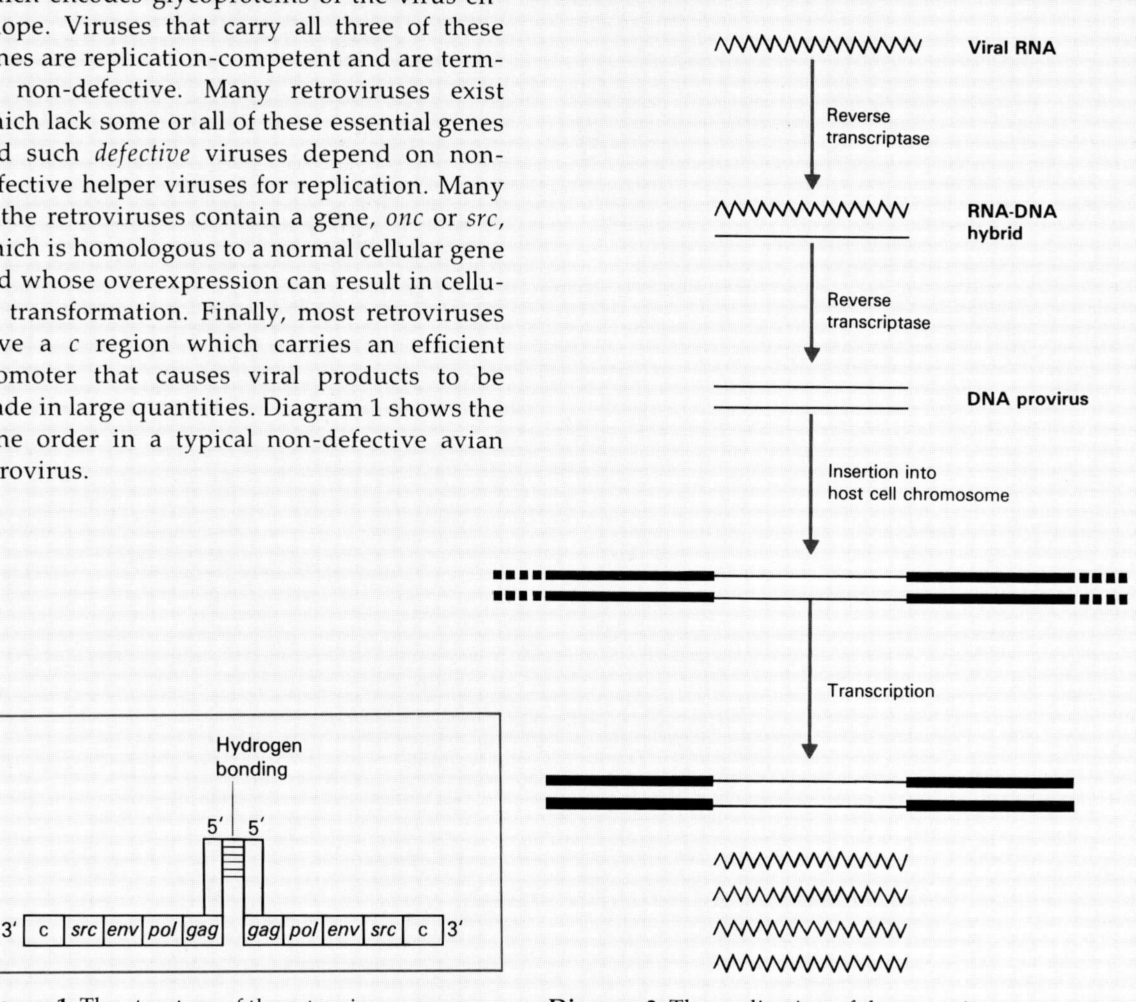

Diagram 1 The structure of the retrovirus genome.

Diagram 2 The replication of the retrovirus genome.

low to be of commercial significance but these will undoubtedly improve in the near future. The fact that none of these substances is marketed as yet is simply a reflection of the long time scale (5–10 years) required for clinical testing and regulatory approval of a drug.

The potential hazards associated with both live, attenuated vaccines and killed vaccines have started a trend towards the generation of subunit vaccines. The basic idea in developing a subunit vaccine is to identify the important immunizing proteins on the virus, clone the relevant genes in a microorganism, purify the viral protein and use it for vaccination. Unfortunately, in most cases the microbially produced protein has low immunizing activity compared with the virus particle. There are three potential reasons for this lack of immunogenicity and all could be related to the fact that bacteria such as *E. coli* do not glycosylate proteins whereas many surface proteins of viruses are glycosylated. First, the carbohydrate itself may be part of an antigenic determinant of the protein. Second, a non-glycosylated protein may express antigenic determinants which are not found on the glycosylated molecule. Third, the attachment of carbohydrate may affect the overall folding of the protein and thus influence antigenic determinants at sites remote from the glycosylated residues.

Since correct glycosylation could be essential for the maintenance of a properly immunogenic molecule, vaccinia virus recombinants have been constructed in which the coding sequence for hepatitis B virus surface antigen (HBsAg) has been placed under the control of a vaccinia promoter. Cells infected with these recombinants not only secrete HBsAg which provokes the formation of antibodies in rabbit, but the protein self-assembles into a structure resembling the DNA-free particles seen in the plasma of chronic carriers of hepatitis. More important, the recombinant virus itself can be used to vaccinate animals. Other viral genes expressed in vaccinia recombinants are listed in Table 9.6. Three of them deserve further comment. First, the recombinant virus expressing the rabies virus glycoprotein G can be administered orally for vaccinating foxes against rabies. Second, the recombinant expressing vesicular stomatitis virus (VSV) nucleoprotein was accidentally injected into a laboratory worker. This individual suffered no adverse effects but developed high antibody titres against the nucleoprotein of VSV. Finally, the recombinant expressing the AIDS virus envelope protein not only immunoprecipitates with antisera from AIDS patients but induces antibody formation when inoculated into mice.

Table 9.6 Viral genes expressed in vaccinia recombinants.

Hepatitis B virus surface antigen
Herpes simplex virus glycoprotein D
Rabies virus glycoprotein G
Vesicular stomatitis virus (VSV) nucleoprotein
Epstein–Barr virus glycoprotein
Influenza virus haemagglutinin
Malaria sporozoite antigen
Human respiratory syncytial virus glycoprotein G
Sindbis virus structural proteins
HTLV-III (AIDS virus) envelope gene

Further reading

GENERAL

Cartwright T. (1987) Isolation and purification of products from animal cells. *Trends in Biotechnology* **5**, 25–30.

Chanock R.M. & Lernes R.A. (1984) *Modern Approaches to Vaccines: Molecular and Chemical Basis of Virus Virulence and Immunogenicity*. Cold Spring Harbor Laboratories, Cold Spring Harbor.

Old R.W. & Primrose S.B. (1985) *Principles of Gene Manipulation: An Introduction to Genetic Engineering*, 3rd edn. Blackwell Scientific Publications, Oxford.

Speir R.E. & Griffiths J.B. (1985) *Animal Cell Biotechnology*, Vols 1 and 2. Academic Press, London.

SPECIFIC

Anson D.S., Austen D.E.G. & Brownlee G.G. (1985) Expression of active factor IX from recombinant DNA clones in mammalian cells. *Nature* **315**, 683–5.

Bermann P.W. & Lasky L.A. (1985) Engineering glycoproteins for use as pharmaceuticals. *Trends in Biotechnology* **3**, 51–3.

Blancou J., Kieny M.P., Lathe R., Lecocq J.P., Pastoret P.P., Soulebot J.P. & Desmettre P. (1986) Oral vaccination of the fox against rabies using a live recombinant vaccinia virus. *Nature* **322**, 373–5.

Chakrabarti S., Robert-Guroff M., Wong-Staal F., Gallo R.C. & Moss B. (1986) Expression of the HTLV-III envelope gene by a recombinant vaccinia virus. *Nature* **320**, 535–7.

Hu S-L., Kosowski S.P. & Dalrymple J.M. (1986) Expression of AIDS virus envelope gene by a recombinant vaccinia virus. *Nature* **320**, 537–40.
Michel M-L., Sobczak E., Malpiece Y., Tiollais P. & Streeck R.E. (1985) Expression of amplified hepatitis B virus surface antigen genes in Chinese hamster ovary cells. *Biotechnology* **3**, 561–6.
Moss B., Smith G.L., Gerin J.L. & Purcell R.H. (1984) Live recombinant vaccinia virus protects chimpanzees against hepatitis B. *Nature* **311**, 67–9.
Roizman B. & Jenkins F.J. (1985) Genetic engineering of novel genomes of large DNA viruses. *Science* **229**, 1208–14.
Smith G.L., Mackett M. & Moss B. (1983) Infectious vaccinia virus recombinants that express hepatitis B surface antigen. *Nature* **302**, 490–5.

10/Monoclonal Antibodies

INTRODUCTION

When a foreign macromolecule (*antigen*) is introduced into the circulatory system of a higher vertebrate it stimulates specific white blood cells, the lymphocytes, to produce antibodies that combine specifically with the macromolecule to facilitate its destruction within, or removal from, the body. The antibodies which are synthesized can be found in the globulin fraction of the proteins that circulate in the blood and hence are called *immunoglobulins*. All immunoglobulin molecules have a similar basic structure consisting of two heavy and two light chains held together by disulphide bonds (Fig. 10.1). There are areas in the heavy and light chains known as *variable regions* in which the amino acid sequence of the protein chain varies from one antibody to another. This variation occurs at the N-terminus of the peptide chains. Each different antibody will therefore have a different amino acid sequence and spatial arrangement and it is conceivable that every possible shape presented by an antigen can be accomodated by some antibody produced by the immune system.

When the immune system of an animal encounters a new antigen it does not respond to the entire surface of the antigen but to specific antigenic determinants (*epitopes*) located on it. Thus a protein antigen may possess several epitopes and would

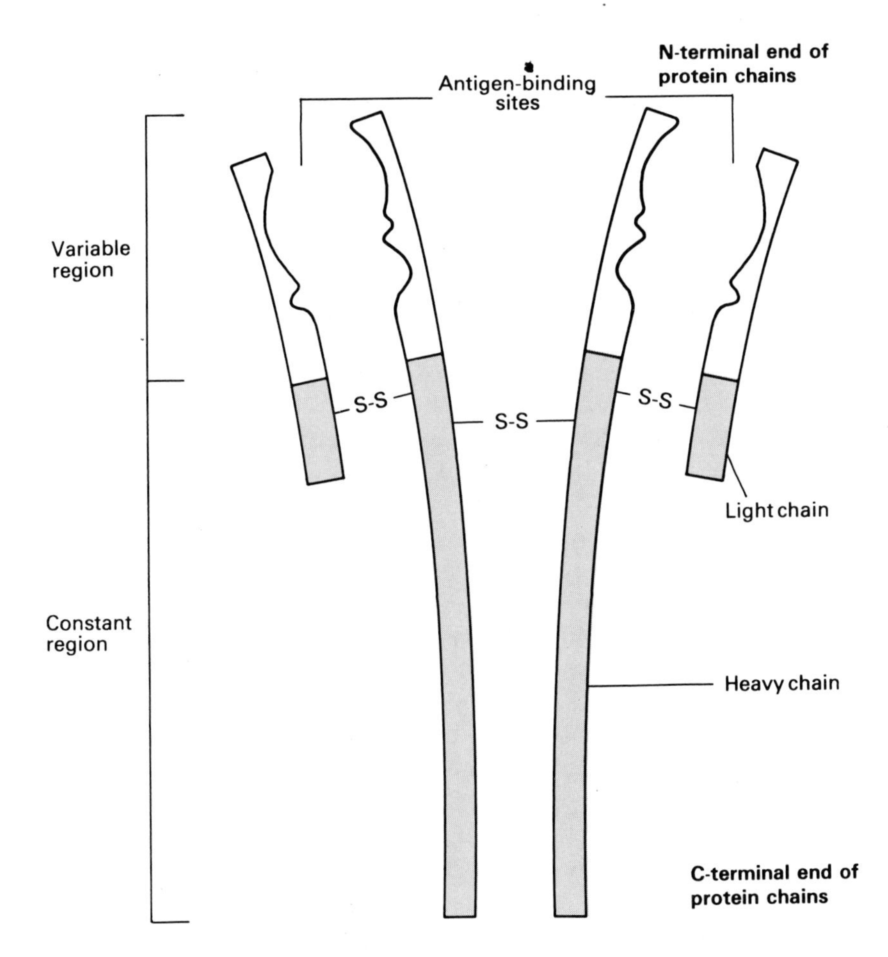

Fig. 10.1 Schematic representation of an immunoglobulin molecule.

induce the formation of several different antibodies; an object the size of a bacterium would possess many different epitopes and induce the synthesis of a correspondingly greater number of antibodies. Following induction of antibody formation the animal can be bled and the serum fraction obtained. However this serum will contain all the different antibodies produced in response to the antigen. Furthermore these antigen-specific antibodies will be diluted with all the other antibodies present in the serum as a result of the animal's previous encounters with other antigens. Such polyclonal antisera have been widely used by biologists but, as will be seen later in this chapter, there are many commercial applications of antibodies where it is essential to have only a single antibody species. These *monoclonal antibodies* can be produced by hybridoma technology.

MONOCLONAL ANTIBODIES FROM HYBRIDOMA TUMOURS

Antibodies are produced by a class of lymphocyte known as B-cells or *plasma cells* (see also Box, p. 131). To produce a monoclonal antibody (MAB) from an immunized animal ideally one would isolate the lymphocytes, culture them *in vitro* and clone out the ones producing the antibody of interest. There are two problems with this approach. First, there has been little success in culturing normal B-lymphocytes by the techniques decribed in Chapter 8. Second, to provide an unlimited supply of antibody, an immortal antibody-producing line is required. Suprising as it may seem, immortal monoclonal antibody-producing cells do exist in nature. They occur in the disease state known as *multiple myeloma*, which is a cancer of the B-lymphocytes. In an individual thus afflicted a neoplastic transformation has occurred in a single plasma cell so that the resulting tumour secretes a homogeneous immunoglobulin. This myeloma protein can comprise 95% of the serum immunoglobulin and the cell which produces it can be cloned easily and grown *in vitro*. The myeloma protein for any one afflicted individual is different from the myeloma proteins of all other individuals and thus different clones producing different monoclonal antibodies can be obtained. However, the choice of monoclonal antibody is restricted to those produced by myelomas studied to date. What is needed is a means of extending the range of antibodies produced by myeloma cells and this was achieved in 1975 in a classic piece of work which later earned a Nobel Prize for George Kohler and Cesar Milstein.

Cells in culture can fuse spontaneously and the frequency with which this occurs can be increased by the addition of fusion-promoting agents such as polyethylene glycol. The immediate fusion products are heterokaryons in which the cell retains both nuclei intact. In many of the heterokaryons the nuclei fuse to generate stable hybrids which express both parental genomes. What Kohler and Milstein did was to fuse normal antibody-producing mouse lymphocytes with a mouse myeloma cell line to generate a *hybridoma*. The hybridoma is derived from a transformed cell and so it has the potential to be grown indefinitely in culture.

At this stage it is important to realize that not just any myeloma cell can be used, only ones which have two particular properties are suitable. First, the myeloma cell line used must itself not be capable of secreting antibody otherwise the hybridoma will secrete a mixture of antibodies. Such cell lines are readily obtained because, although in most cases of multiple myeloma there is an overproduction of a particular immunoglobulin, there are instances where a particular lymphocyte multiplies uncontrollably without antibody production. Second, the myeloma cells used for fusion are specially selected by growing them in the presence of 8-azaguanine. Most of the cells are killed by this technique but a few survive and these resistant cells have a defect in the enzyme hypoxanthine phosphoribosyl transferase (HPRT).

When HPRT-negative cells are grown in a mixture of hypoxanthine, aminopterin and thymidine (HAT medium) the cells will die because they can no longer synthesize DNA (Fig. 10.2). The aminopterin blocks the main pathway for purine and pyrimidine biosynthesis. Normal cells can use the thymidine to make pyrimidines and, using HPRT, can convert the hypoxanthine to purines. In a fusion mixture of lymphocytes and myeloma cells neither cell can grow in the HAT medium. The lymphocytes will survive for a few days before dying out, as will the myeloma cell, because they lack HPRT. However, the hybridoma cells will possess the ability of the myeloma cells to grow *in vitro* and the normal HPRT gene from the lymphocytes with which they

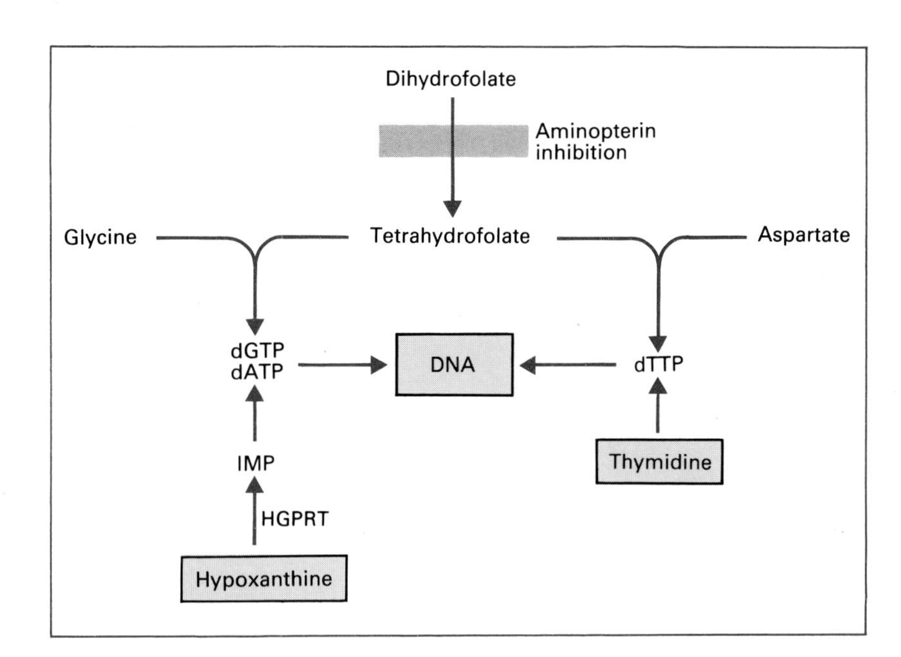

Fig. 10.2 The selective basis for the use of HAT medium. See text for details.

have fused. Thus the hybridoma will grow successfully even in the HAT selection medium.

MAKING A HYBRIDOMA

The starting point for making a hybridoma which secretes the desired MAB is to immunize an animal, e.g. a mouse, with the appropriate antigen. Normally the antigen is injected subcutaneously or into the peritoneal cavity along with an adjuvant to stimulate the immune system. The animal is injected on several occasions and with each successive immunization there is increased stimulation of the B-lymphocytes which are responding to the antigen. The final dose of antigen is given intravenously three days before the animal is killed. The intravenous injection ensures a high dose of antigen and three days after immunization the immune-stimulated cells will be growing maximally. This simplifies the selection process later.

After the immunized animal has been killed its spleen is aseptically removed and gently disrupted to release the lymphocytes and red blood cells. The lymphocytes are separated from the red blood cells and splenic fluid by density gradient centrifugation. After washing, the lymphocytes are mixed with an HPRT-negative myeloma cell line. The mixture of cells is exposed to the fusion-promoting agent polyethylene glycol, but only for a few minutes since it is cytotoxic. The cells are then washed free of the polyethylene glycol by suspending the pellets in fresh media. The washed cells comprise a mixture of hybridomas, unfused myeloma cells and unfused lymphocytes. By using HAT growing medium the hybridomas can be selected as described above.

If all the hybridomas that occurred after a fusion were grown together, then a polyclonal antibody mixture would be obtained. Consequently, single antibody producing hybridoma cells need to be isolated and grown on individually. This is done by diluting a suspension of hybridoma cells to such an extent that individual aliquots contain, on average, only one cell. The cells are then grown in fresh medium. The problem here is that animal cells do not readily grow in isolation but ways of overcoming this are described in Chapter 8 (see p. 102). Each clone is then examined to determine if it produces the desired MAB. Here one encounters a problem of logistics. Even with the best immunization schedule less than 5% of the splenocytes will be producing antibodies and, of those that are, only a small proportion will have responded to the antigen used for immunization. Consequently a mass screening procedure needs to be undertaken and success is not guaranteed! The various steps involved in making a hybridoma are summarized in Fig. 10.3.

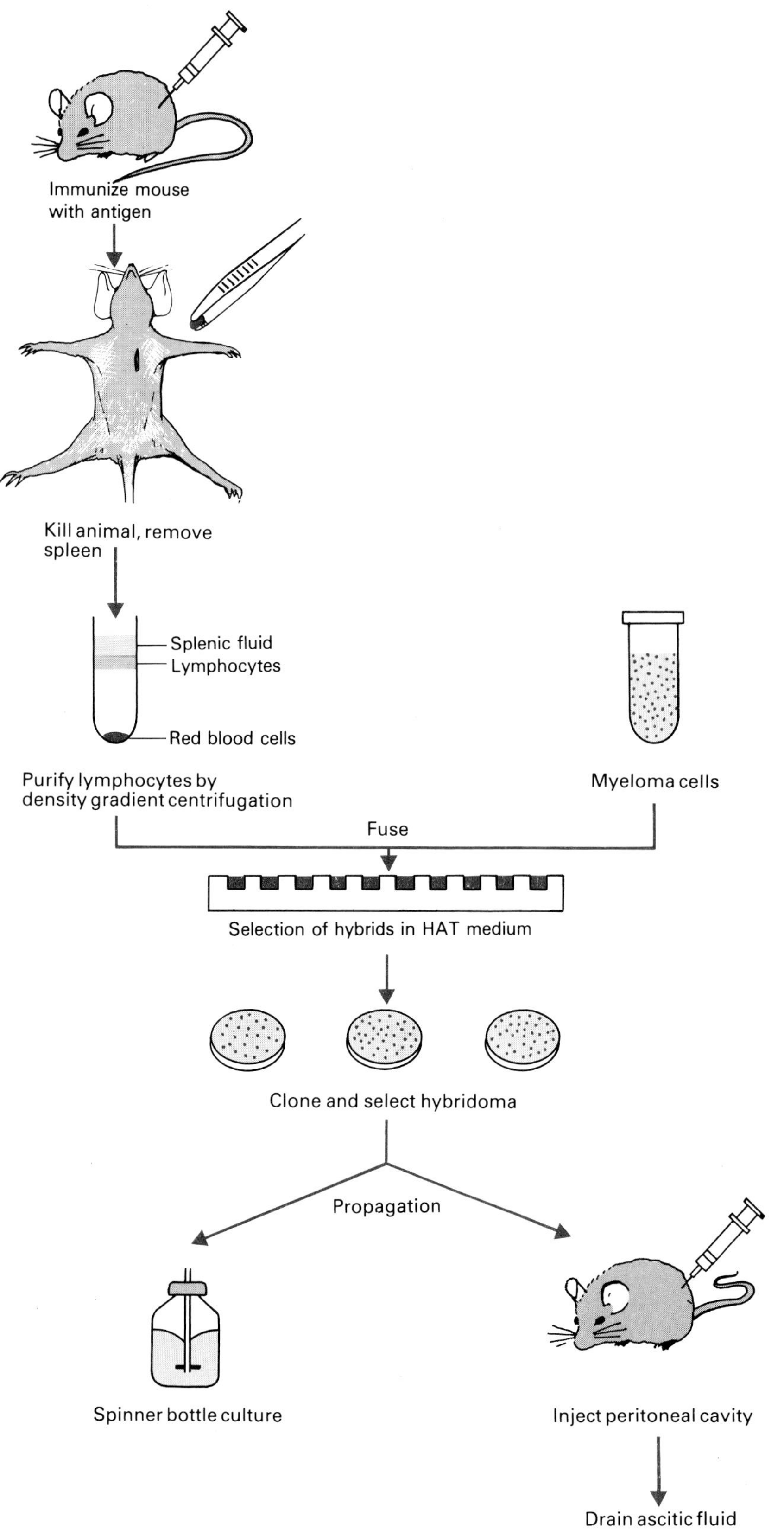

Fig. 10.3 Steps in the production of a hybridoma.

LARGE-SCALE PRODUCTION OF MABS

Once the correct hybridoma has been selected, it can be stored frozen and cultured whenever required. Since the hybridoma is a transformed cell line it grows readily in culture but the antibody titre is low. In roller bottle culture antibody levels of 5–10 mg/l of culture medium are not uncommon. One solution is to grow the hybridoma to high cell density in suspension culture with or without the use of microcarriers. In this way antibody levels of 10–100 mg/l can be obtained. An interesting method patented by the Damon Corporation is to grow the hybridomas inside hollow microspheres which are surrounded by a porous membrane (Fig. 10.4). Protected inside the microsphere the cells multiply and grow to higher cell densities than is possible with any other culture system. Since the antibody is retained by the membrane it does not become contaminated with other immunoglobulins if serum has to be included in the growth medium. Using this microencapsulation technology 100 mg to 1 g of

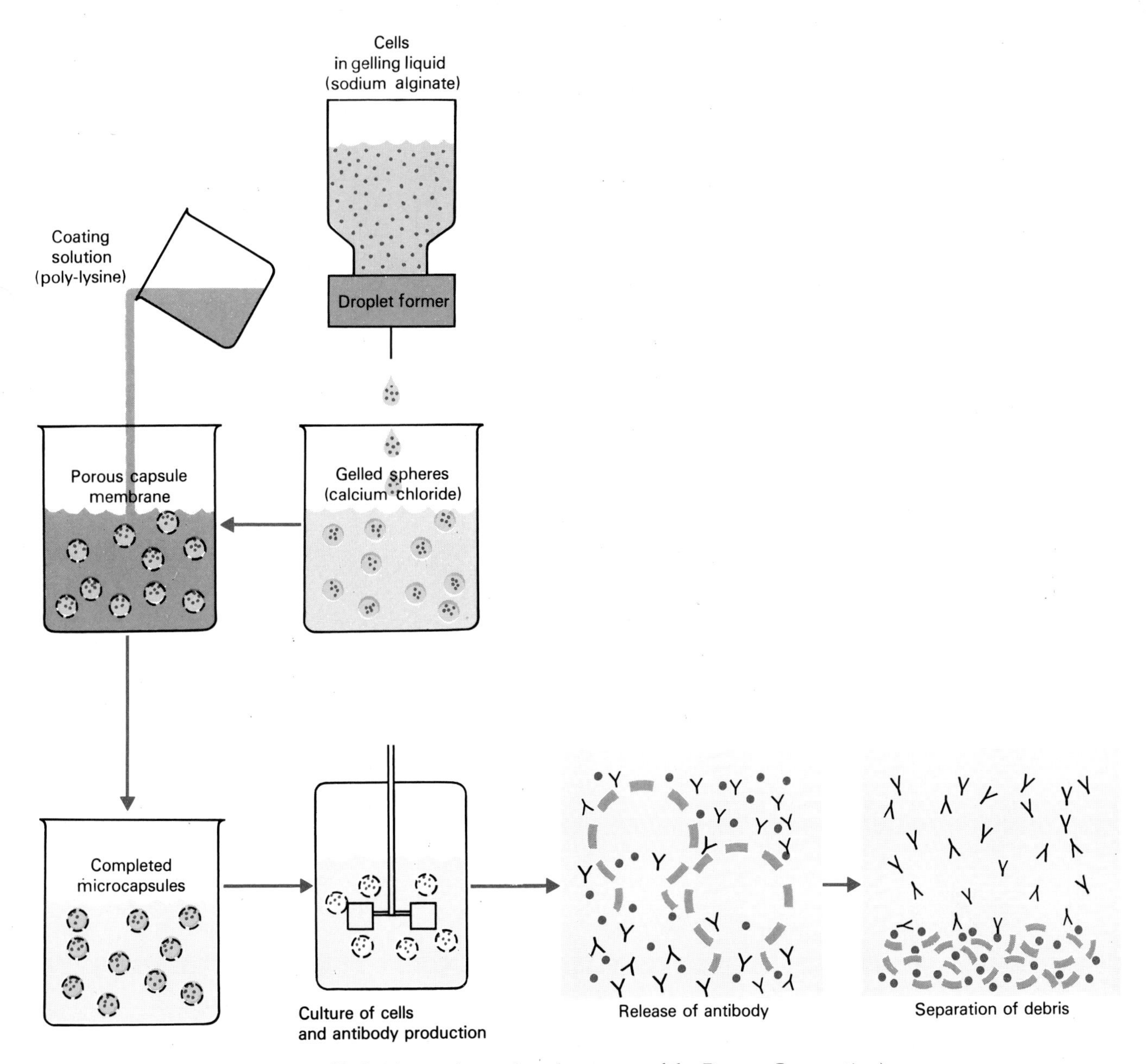

Fig. 10.4 Microencapsulation of hybridomas (reproduced courtesy of the Damon Corporation).

antibody can be produced per litre of culture medium and this makes large-scale production of MABs commercially feasible.

An alternative way of obtaining a high concentration of antibody is to grow the hybridoma as ascites, i.e. as a suspension of tumour cells within the peritoneal cavity of an isogenic species. The ascitic fluid which forms following injection of the tumour can contain as much as 20 g/l of MAB, a level superior to that obtained by the best in-vitro cultivation methods. Some commercial suppliers of MABs use the ascites route but it has its disadvantages. Not only is it labour intensive, but specific pathogen-free (SPF) animals need to be used and kept quarantined at all times to prevent contamination of the MAB. This adds greatly to the cost. In addition, why sacrifice large numbers of laboratory animals when an alternative is available?

Whatever the method used to grow the hybridoma, the MAB produced usually is purified prior to use. If the hybridoma has been grown *in vitro*, purification of the MAB will be facilitated if the culture medium contained only serum albumin and not complete serum. After precipitation with ammonium sulphate the MAB can be purified by ion exchange chromatography, e.g. on a DEAE-based matrix (Fig. 10.5).

HAZARDS ASSOCIATED WITH MABS

The majority of mammalian cells contain in their chromosomes genetic information related to retro-

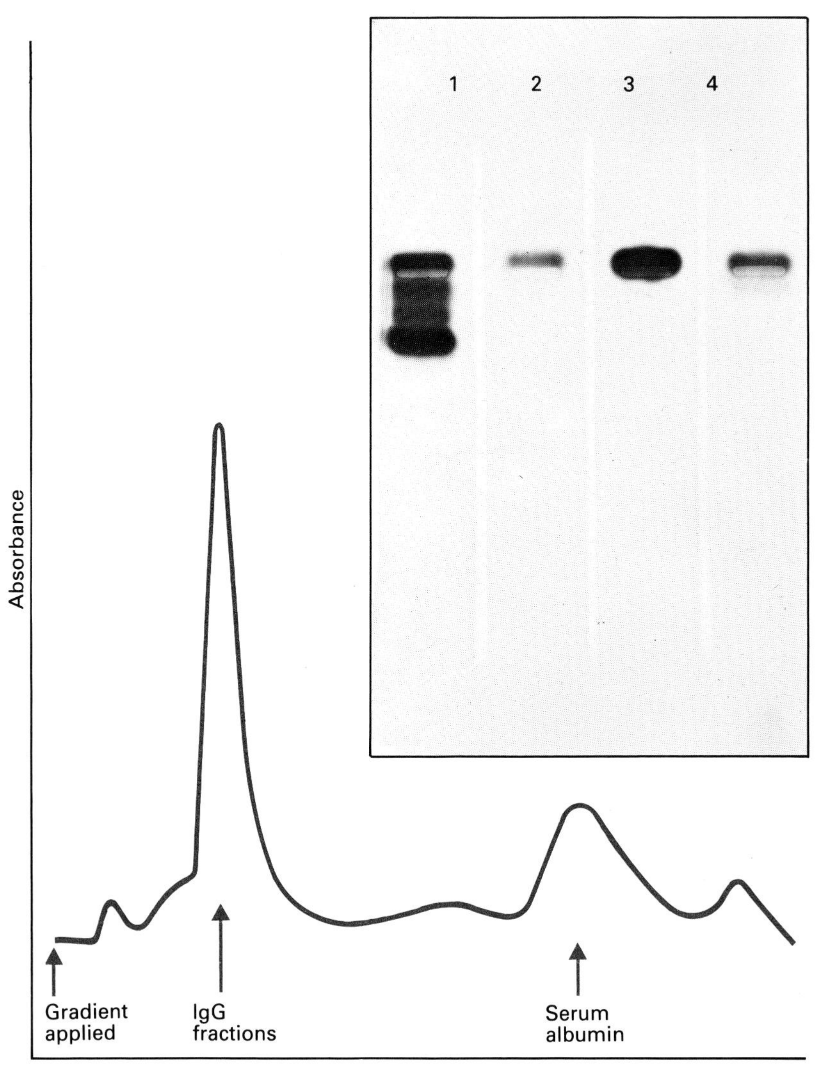

Fig. 10.5 Purification of a MAB from ascites fluid. The main figure shows the separation of IgG from contaminating proteins by ion exchange chromatography of ascitic fluid. The inset shows an electropherogram of crude ascitic fluid (track 1) and various IgG fractions (tracks 2–4).

viruses. Mice, and in particular the inbred strains used in laboratory studies, carry a number of different types of these viruses and murine cell lines frequently release virus. Retrovirus production has been noted in a number of myeloma cell lines and in hybridomas prepared from them. It is possible to find myeloma lines that do not actively produce virus and, although their use does not guarantee that a hybridoma will be virus free, it does reduce the risk of contamination. Due to the dangers which those viruses can pose, regulatory authorities have stipulated that medicinal products containing MABs or produced using MABs will not be approved unless the MABs and/or the hybridomas can be shown to be virus free. Table 10.1 lists those viruses of concern to the US Food and Drug Administration: demonstrating their absence is not easy! An alternative method of producing MABs, and one which eliminates the virological hazards, is to synthesize them in bacteria using recombinant DNA technology. This has been achieved recently but, as yet, the concentrations are too low to be commercially viable.

Table 10.1 Viruses which are potential contaminants of MABs produced by hybridomas of murine origin.

Viruses
Adenovirus
Encephalomyelitis virus
Encephalomyocarditis virus
Lymphocytic choriomeningitis virus
Hepatitis virus
Polyoma
Retroviruses
Sendai virus
Thymic virus
Reovirus
Haemorrhagic fever virus
Ectromelia virus
Lactate dehydrogenase virus
Cytomegalovirus

Applications of MABs

For many years polyclonal antisera produced by bleeding immunized animals have been widely used for diagnostic purposes, particularly in clinical microbiology and experimental biology. Monoclonal antibodies are far more versatile because of their specificity and high purity and hence they are used in such diverse applications as human therapy and commercial protein purification. As a consequence a whole new industry has grown up and in recent years several biotechnology companies have been founded whose income comes mostly from the sale of MABs.

As reference to Table 10.2 will show, a detailed description of all the uses of MABs is outside the scope of this book. A few representative examples

Table 10.2 Some of the applications of MABs.

Improved sensitivity and reproducibility of existing immunoassays or new assays for:

Histocompatibility antigens	Interleukins	Blood clotting factors
Fibronectin	Complement components	Oestrogen
Blood group antigens	Interferons	Human growth hormone
Sperm antigens	Progesterone	
	Gastrin	

Diagnosis of:

Sexually transmitted diseases
Cancer (by detection of oncofetal antigens)

Therapy:

Correction of drug overdose
Reduction of risks associated with bone marrow transplants
Detection of tumour metastases
Treatment of cancer (directly or by targetting cytotoxic drugs)

are given below and those wishing further information should consult the monograph by Sikora and Smedley (1984).

IMPROVED DIAGNOSTIC REAGENTS

Antibody–antigen interactions are used widely for the assay of a wide range of compounds including hormones, antibiotics and biological effector molecules such as interferons and blood clotting factors. Also, the ability of antibodies to agglutinate or precipitate cells has long been used in diagnostic microbiology and in blood typing. Prior to the development of MABs polyclonal antisera were used but these have a number of disadvantages. For example, serum batches from different animals or groups of animal, or even one animal at different times, have different antibody contents and usually do not show an identical reaction pattern. Thus there will be considerable batch to batch variation in the diagnostic reagent and this generates a requirement for repeated calibration. This inherent variability does not occur with MABs since they should be produced in a consistent fashion by repeated subculture of a master cell bank of a hybridoma. Another problem is that polyclonal antisera contain only a small proportion of the antibody of interest. Since the rate of interaction between an antigen and an antibody is concentration dependent this means that reaction times are prolonged. The homogeneity of MABs reduces reaction times and there is less likelihood of non-specific cross-reactions occurring. As a consequence of these advantages — consistency, specificity and speed — a blood grouping kit based on MABs is now on the market. For the same reasons it is also possible to diagnose pregnancy or ovulation more precisely.

PROTEIN PURIFICATION

Monoclonal antibody affinity columns are readily prepared by coupling MABs to a cyanogen bromide-activated chromatography matrix, e.g. Sepharose. MABs immobilized in this way are particularly valuable for the purification of proteins. There are at least three advantages of immunoaffinity methods over conventional purification techniques. First, since the MAB has a unique specificity for the desired protein the level of contamination by unwanted protein species usually is very low. Second, since the MAB antigen complex has a single binding affinity it is possible to elute the required protein (antigen) in a single, sharp peak (Fig. 10.6). Third, the concentration of the required protein relative to total protein in a mixture is irrelevant since the capacity of the antibody remains unchanged. The data in Table 10.3 show that 97% recovery of interferon was obtained by immunoaffinity chromatography when the starting concentration was less than 0.02%. The fact that a 5300-fold purification was obtained in a single step highlights the power of the method and it is not surprising that it is being used commercially to purify recombinant-derived interferon-α_2. However, a word of caution is necessary: 100% pure protein is difficult to achieve by this method because there is always a tendency for small amounts of immunoglobulin to leak off the immunoaffinity column. Furthermore, MABs often do not distinguish between intact protein molecules and fragments containing the antigenic site.

BONE MARROW AND ORGAN TRANSPLANTATION

There are several diseases where the injection of

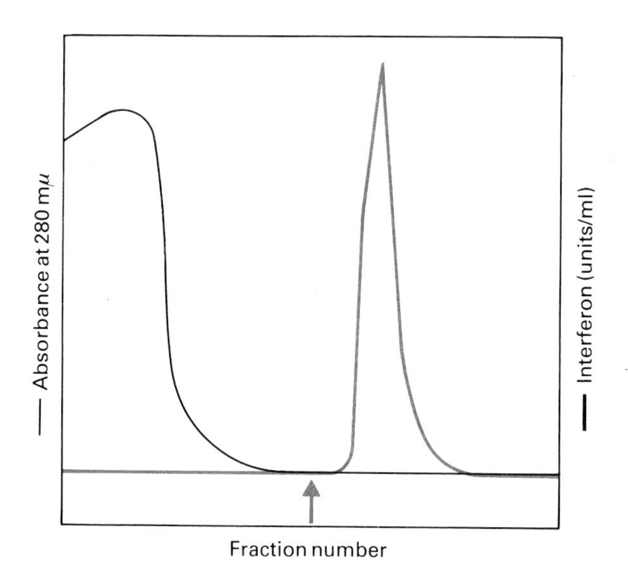

Fig. 10.6 Purification of interferon-α by affinity chromatography. Crude interferon was loaded onto a column of anti-interferon MAB immobilized to Sepharose. After loading the unbound protein was eluted by washing with buffer. At the point shown by the arrow, an acid-wash (pH 2.0) was started to elute the interferon.

Table 10.3 Purification of crude interferon from tissue culture medium using an immobilized anti-interferon antibody (adapted from Secher & Burke 1980).

	Before	After
	Immunoaffinity chromatography	
Volume of interferon-containing solution	100 ml	0.5 ml
Interferon titre (u/ml)	7.2×10^4	1.5×10^7
Total units	7.2×10^6	7.0×10^6
Specific activity of interferon (u/mg)	3.3×10^4	1.8×10^8
Purification factor	1	5300
Recovery of interferon	100	97
Size of immunoaffinity columns	0.5 ml	

bone marrow from a healthy individual may be of considerable benefit to a patient. These include a wide variety of congenital marrow aplasias where erythrocyte stem cells stop dividing, resulting in severe anaemia. These aplasias can be cured if the red cells in the marrow can be replaced by normal stem cells. Disorders of leucocyte and lymphocyte stem cells also occur resulting in greatly decreased resistance to microbial infection.

There are two major problems associated with bone marrow transplantation. The first of these is the rejection of the donor cells by the host. Eventually the use of MABs will permit better typing of donor and recipient lymphocytes to enable better cross-matching. In the meantime, immunosuppressive therapy is required for a prolonged period of time after transplantation but there is a consequently higher risk of mortality from Gram-negative bacteraemia. The problem stems from the fact that even if antibiotics successfully combat the infection, the released endotoxin can still cause death. A polyclonal anti-endotoxin antibody has been produced in humans and shown to be effective in reducing the death from bacteraemia. More recently a human MAB against *E. coli* endotoxin has been produced which protects mice against lethal Gram-negative bacteraemia. Results from the use of this MAB in humans are awaited with interest; not only might the increased purity of the MAB mean increased reactivity but, provided care has been taken, the MAB should be free from hepatitis, cytomegalovirus and AIDS, which are potential contaminants of human polyclonal antisera.

A second problem associated with marrow transplants is that the marrow contains donor T-lymphocytes. These can recognize cells in the new host as being foreign and begin to destroy them. This produces the clinical syndrome known as graft-versus-host disease (GVH), which can lead to death although it is partly controlled by immunosuppressive drugs. An anti-T-cell monoclonal antibody is available and has been used to remove T-cells from the donor marrow prior to transplantation, leading to a reduction in GVH disease. A second use for the anti-T-cell monoclonal antibody, which has received regulatory approval, is to prevent rejection of donor organs such as heart, liver and kidneys after transplantation. Following administration of the antibody the T-cells are removed from the circulation and the patient temporarily becomes immunologically incompetent. A foreign organ, such as a kidney introduced into the body at this time, is not recognized as being antigenic, so antibodies to it are not synthesized when T-cells reappear in the bloodstream.

IMMUNOTOXINS AS THERAPEUTIC AGENTS

Many tumour cells possess additional or new antigens on their surface in comparison to the normal cells from which they are derived. If MABs which bind to these tumour cells are isolated, it should be possible to isolate one or more tumour-specific MAB. Such antibodies may form the basis of a new form of tumour chemotherapy: immunotoxicology.

Ricin is a protein derived from the castor oil plant and which is cytotoxic. It consists of two peptide chains linked by a disulphide bond. The B-chain binds to galactose residues on the surface of sensitive cells and this binding facilitates the entry of the

A-chain. Once the A-chain gains access to the cell cytoplasm it enzymatically and irreversibly alters the ribosomes so that they can no longer function. This leads to cell death. The toxic A-chain of ricin can be separated from the B-chain and coupled to the MAB. This newly created immunotoxin should bind to tumour cells and not normal cells (Fig. 10.7) and lead to death of the former. Since the A-chain of ricin inhibits protein synthesis and not cell division it kills even those cells which are potentially harmful but not actively dividing. However, such immunotoxins must be chemically stable in the body; fortunately, studies to date indicate that immunotoxins injected into the body are stable.

An alternative approach is to couple the MAB to a radiosotope such that the tumour cell is killed by

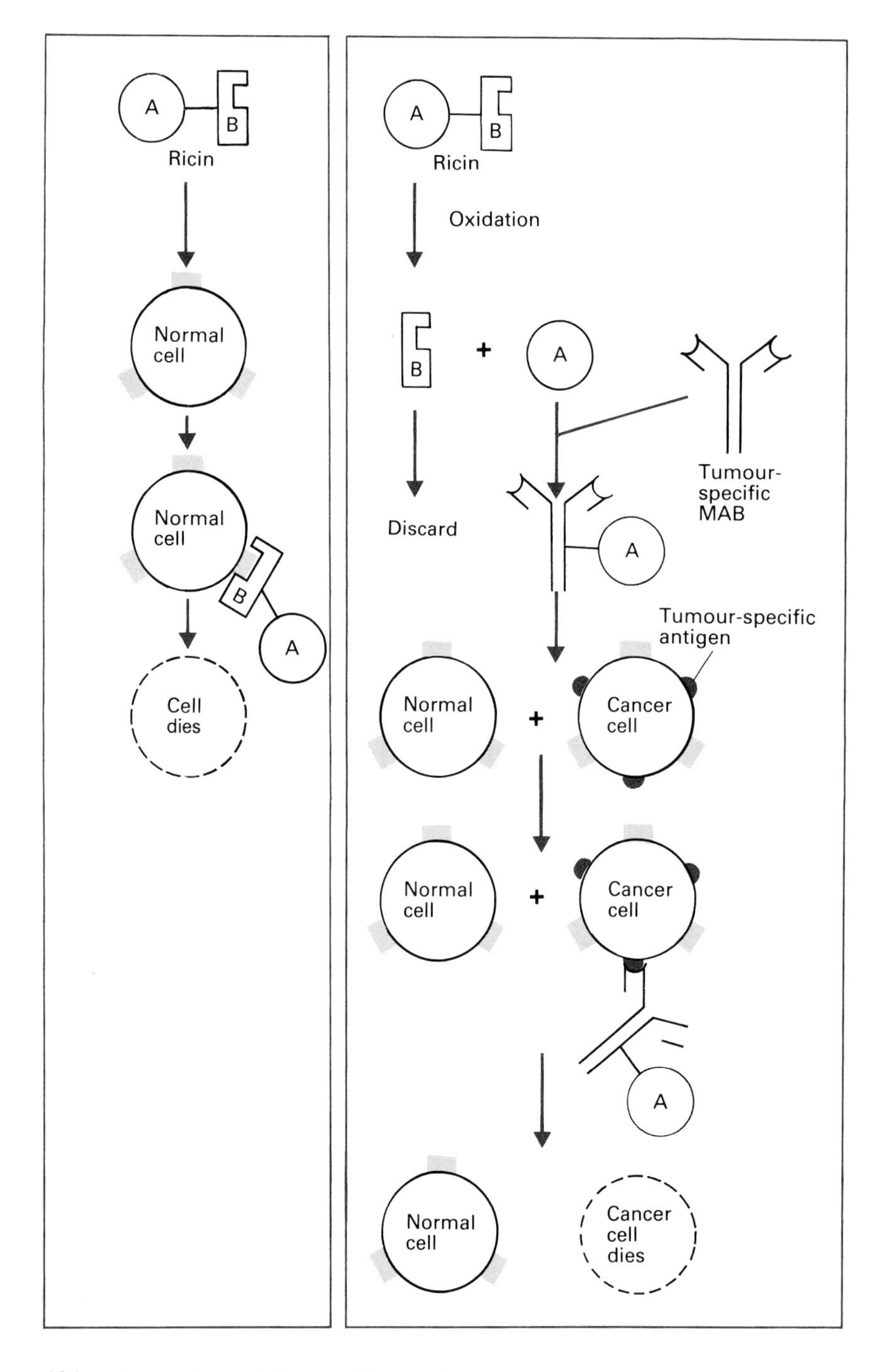

Fig. 10.7 Use of immunotoxins as therapeutic agents.

irradiation. For this purpose the radioisotope selected needs to have high energy (to kill target cells), low penetration (to prevent damage to surrounding tissue), short half-life (so that the patient is not a radiation hazard!) and the products of radioactive decay should be inert. One isotope which appears suitable and which can be coupled to MABs is astatine whose half-life is 13 h. Unfortunately, astatine has a propensity to be incorporated into bone and this is not desirable. An alternative is yttrium-90 which has a half-life of 64 h. Although yttrium-90 is not incorporated into bone the β particles which it emits have greater tissue penetration than the α particles emitted by astatine.

CHIMAERIC ANTIBODIES

The genetics of antibody synthesis are very complex since both the heavy and light chains (Fig. 10.1) are encoded by multiple DNA segments which undergo complex rearrangements in the intact animal. In simplistic terms, four protein-coding regions are required, as shown in Fig. 10.8. The genes for rodent and human heavy chain variable (V_H), heavy chain constant (C_H), light chain variable (V_L) and light chain constant (C_L) regions have been cloned and expressed in bacteria. More important, these cloned genes can be introduced into myeloma cells where they direct the formation of the appropriate proteins. By first recombining various genes *in vitro*, e.g. by coupling human C_H and C_L genes with mouse V_H and V_L genes, and then introducing them into myeloma cells it is possible to obtain novel hybrid (chimaeric) antibodies which are partly human and partly rodent. Such combinations may be more therapeutically useful than MABs of purely murine origin. In the same way it is possible to couple foreign proteins such as the ricin A gene to all or part of the heavy chain.

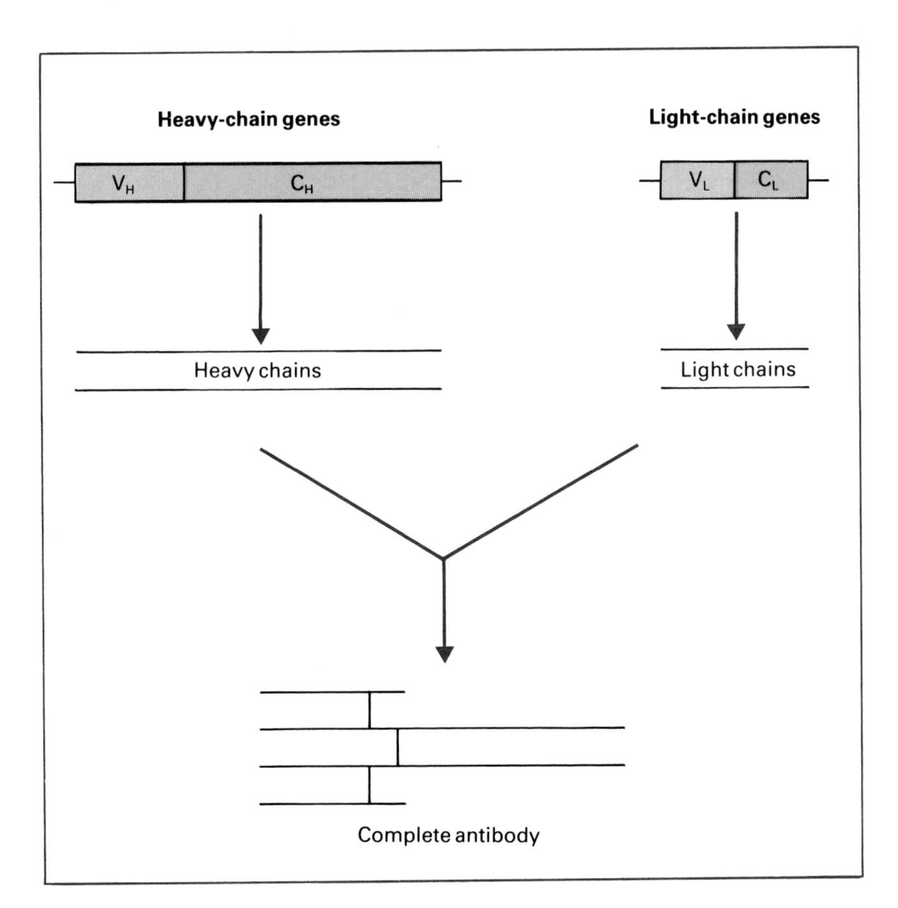

Fig. 10.8 Simplified representation of the genetics of antibody formation.

Further reading

GENERAL

Oi V.T. & Morrison S.L. (1986) Chimaeric antibodies. *Biotechniques* **4**, 214–21.

Seiler F.S., Gronski P., Kurrle R., Lueben G., Harthus H.P., Ax W., Bosslet K. & Schwick G. (1985) Monoclonal antibodies: their chemistry, functions and possible uses. *Angewandte Chemie* (International edition) **24**, 139–60.

Sikora K. & Smedley H.M. (1984) *Monoclonal Antibodies*. Blackwell Scientific Publications, Oxford.

SPECIFIC

Bankert R.B. (1984) Development and use of monoclonal antibodies in the treatment of cancer. *Cancer Drug Delivery* **1**, 251–67.

Benjamin R.J. & Waldmann H. (1986) Induction of tolerance by monoclonal antibody therapy. *Nature* **320**, 449–51.

Filipoitch A.H., Ramsay N.K. & Warkentin P.I. (1982) Pretreatment of donor bone marrow with a monoclonal antibody OKT3 for prevention of graft versus host disease. *Lancet* **i**, 1226–9.

Klausner A. (1986) Taking aim at cancer with monoclonal antibodies. *Bio/technology* **4**, 185–94.

Kohler G. & Milstein C. (1975) Continuous cultures of fused cells producing antibodies of predefined specificity. *Nature* **256**, 495–7.

Kozak R.W., Waldmann T.A., Atcher R.W. & Gansow O.A. (1986) Radionuclide-conjugated monoclonal antibodies: a synthesis of immunology, inorganic chemistry and nuclear science. *Trends in Biotechnology* **4**, 259–64.

Secher D.S. & Burke D.C. (1980) A monoclonal antibody for large scale purification of luman leukocyte interferon. *Nature* **285**, 446–50.

Teng N.N.H., Kaplan H.S., Herbert J.M., Moore C., Douglas H., Wunderlich A. & Braude A.I. (1985) Protection against Gram-negative bacteraemia and endotoxemia with human monoclonal IgM antibodies. *Proceedings of the National Academy of Sciences, USA* **82**, 1790–4.

11/Transgenic Animals and Human Gene Therapy

INTRODUCTION

Gene therapy is the correction of an inborn error of metabolism by the insertion into the afflicted organism of a normal gene. Ideally, the inserted gene will be correctly targeted and regulated. This has been achieved in a number of experimental animal systems and it will not be long before it will be possible to cure some human diseases in a similar way. This immediately raises a whole series of ethical questions, some of which are discussed later (see p. 169), but suffice it to say that clinical protocols are being prepared.

In the case of animals other than man there is no overwhelming desire to keep alive those individuals with severe genetic impairment. However, the same techniques which are being used for gene therapy could be developed into sophisticated animal breeding programmes either to:

1 introduce new characteristics not available before;

2 or to introduce existing characteristics more quickly than by conventional breeding.

As far as contemporary thought is concerned, the insertion of genetic material into a human being for the sole purpose of correcting a genetic defect in that patient, i.e. somatic cell gene therapy, is socially acceptable. Attempts to modify germ cells or to enhance or 'improve' a 'normal' person by gene manipulation are abhorrent, particularly so in view of the difficulties which would be encountered using current techniques. By contrast, in animal breeding, modification of germ cells and 'improvement' of the animal are the desired goals. As will be seen later, the modification of germ cells by gene manipulation *in vitro* is possible and this process is given the name *transgenesis*. Transgenesis is not restricted to animals but can also be achieved in plants (see Chapter 13).

GENE DELIVERY METHODS

Regardless of whether somatic or germ cells are to be modified, a method of introducing exogenous DNA into those cells is required. Table 11.1 lists the methods currently available. The first two methods have been dealt with in detail in Chapter 9 and will not be described again here. Microinjection is a widely used technique but is a laborious procedure as only one cell at a time can be injected; it also has a high failure rate (see for example Table 11.2) because even in the hands of experienced operators many of the eggs are damaged. Electroporation and cell fusion are not used much because little is known about the mechanics of the processes but they would find favour if they could be shown not to result in damage to the recipient cells.

None of the delivery methods listed in Table 11.1 fulfils the criteria of being able to transfect the majority of cells with the minimum cellular damage/mortality. Where somatic cells are to be modified the use of viral vectors or calcium phosphate-mediated uptake is indicated; for modification of germ cells microinjection is the method of choice. Whatever the DNA delivery method used one problem still remains: there is no control over where the donor DNA will integrate in the recipient genome and whether this will result in any pathological condition.

PRODUCTION OF TRANSGENIC MICE

As indicated above, the aim of transgenesis is to

Table 11.1 Methods of introducing exogenous DNA into mammalian cells.

1 Use of viral vectors
2 Calcium phosphate-mediated DNA uptake
3 Microinjection of eggs
4 Fusion of DNA-loaded membranous vesicles (e.g. liposomes) to target cells
5 Electroporation (DNA uptake mediated by electric current)

produce animals with a hereditable change in their genotype such that the benefits of gene manipulation can be passed to their offspring. Economically the current interest is in the modification of farm animals but in the future attention may be paid to other food sources such as fish. With mammals, the technique used is the microinjection of DNA into fertilized eggs followed by implantation of the manipulated ova in a foster mother. For *experimental* purposes the mouse is a particularly appropriate choice of animal: a super-ovulated mouse can yield up to 40 eggs, reimplantation is relatively easy and a mouse can carry up to 20 offspring.

In practice, about 2 pl of buffer containing cloned plasmid DNA is injected into one of the pronuclei of the newly fertilized egg (Fig. 11.1). The male pronucleus, derived from the sperm, is larger than the female pronucleus and so is chosen for injection. The two pronuclei subsequently fuse to form the diploid zygote nucleus of the fertilized egg. The injected embryos are cultured *in vitro* until they have undergone a number of divisions and are then implanted in pseudopregnant foster mothers. Between 3% and 40% of the animals developing from these embryos contain copies of the exogenous DNA integrated into their chromosome. In any particular experiment the copy number of the integrated plasmid sequence in individual transgenic animals can vary over a hundredfold range and the chromosomal location differs.

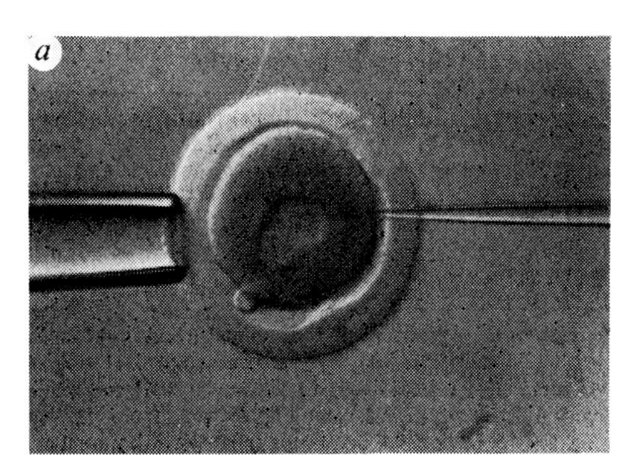

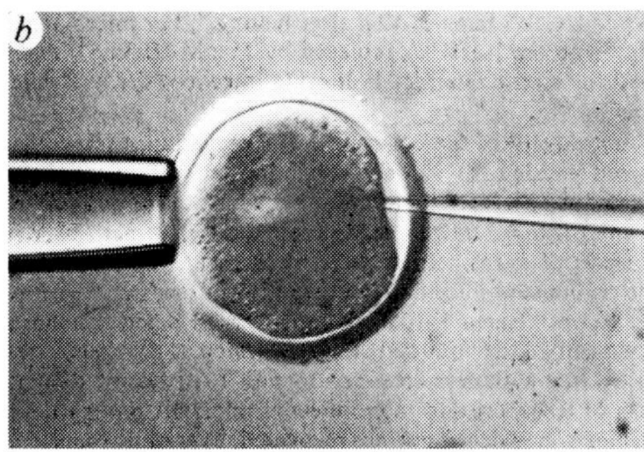

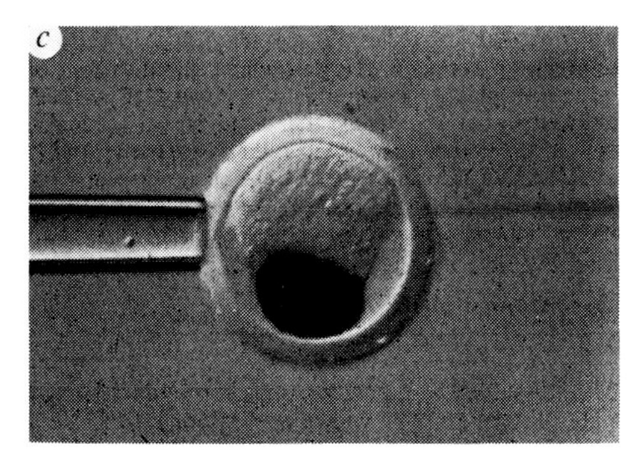

Fig. 11.1 Injection of DNA into the pronucleus of a newly fertilized egg.

Most experiments on transgenesis have made use of the mouse metallothionein (MMT) gene promoter. MMT encodes a small cysteine-rich polypeptide which is involved in heavy metal homeostasis. As such it is inducible not only by heavy metals but also by glucocorticoid hormones. Although distributed in many tissues it is most abundant in the liver. By coupling the MMT promoter to a Herpes Simplex thymidine kinase (TK) gene it is possible to raise transgenic mice in which TK activity shows a tissue distribution similar to MMT and is inducible by cadmium ions. This experiment shows that it is possible for a foreign gene, in this case from a virus, to be integrated into the genome of a mouse. More important, the TK gene is inherited by offspring as though it were integrated into a single chromosome. However, the level of expression in the offspring varies considerably from that of the parent, some progeny having increased expression and others decreased expression.

A more spectacular example is the coupling of the MMT promoter to a rat growth hormone gene. Transgenic mice carrying this gene construct grow significantly faster and larger than their litter-mates which do not (Figs 11.2 and 11.3). In some cases the transgenic mice grow to almost double the weight of normal mice even when the recipient mice are homozygous for growth hormone deficiency. Again, the donor gene is inherited in a Mendelian fashion.

In an ideal transgenesis experiment the promoter selected would be one which enables the normal cellular control patterns to be exerted on the gene of interest, i.e. the gene should be switched on and off at the correct time and in the correct tissues. Success of this kind has been achieved with transgenic mice

carrying the human insulin gene. Not on
gene expression restricted to the islet cells
mouse pancreas but serum levels of human i
were properly regulated by glucose, aminc
and an oral hypoglycaemic agent (tolbutami(

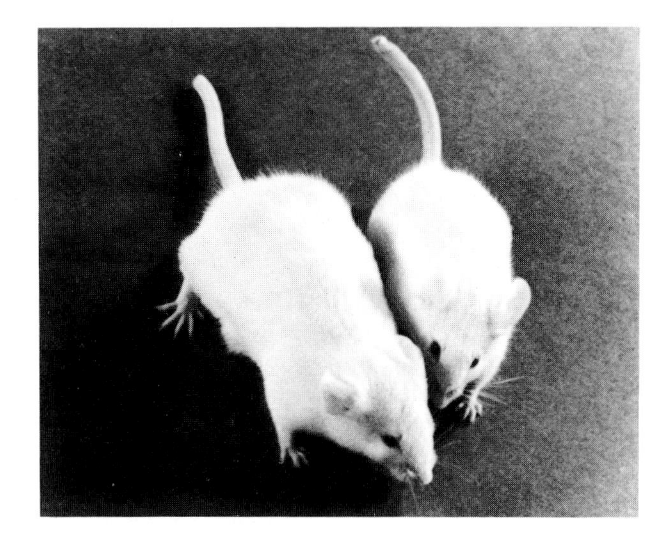

Fig. 11.2 Two male mice from transgenesis experiments. The two mice are siblings and approximately ten weeks old. The mouse on the left contains a new gene composed of the MMT promoter fused to the rat growth hormone structural gene. The male with the new gene weighs 44 g and his sibling without the gene weighs 29 g. The gene is passed on to offspring which also grow larger than controls. In general, mice that express the gene grow 2–3 times as fast as controls and reach a size up to twice normal. (Photograph courtesy of Dr R. Brinster, University of Pennsylvania.)

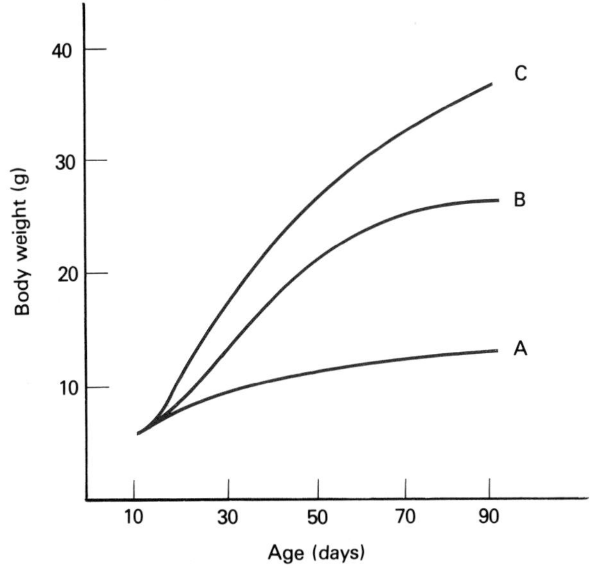

Fig. 11.3 Effect of rat growth hormone on growth of mice. Curve A shows the growth of mice homozygous of a defect resulting in the absence of murine growth hormone. The growth of transgenic litter-mates carrying multiple copies of a rat growth hormone is shown in curve C. Note that the rate of growth and final size of the transgenic mice is greater than that of wild-type mice (curve B). (Adapted from Hammer *et al.* 1984.)

TRANSGENESIS IN LARGE MAMMALS

While transgenic mice are of interest as an experimental tool, the technique needs to be applied to farm animals if any commercial benefit is to ensue. Working with large domestic animals is much more difficult than with mice. First, they do not produce as many eggs, e.g. a ewe on average only gives three or four. Second, reimplantation of manipulated embryos is more difficult because sheep and cattle will not give birth to more than two offspring. Finally, the eggs of many domestic animals have such an opaque cytoplasm that it is impossible to see their pronuclei or nuclei without resorting to special techniques. One technique which has been used to visualize the pronucleus in pig embryos is centrifugation (see Fig. 11.1) to separate the cytoplasmic contents. This centrifugation step does not appear to affect their subsequent development. The difficulties cited above are evident from the data in Table 11.2 which show the results obtained when attempts were made to isolate transgenic rabbits, sheep and pigs.

So far, the enhanced growth of mice after the transfer of the human growth hormone gene is an effect that has not been duplicated in other animals. For example, pigs expressing high levels of growth hormone do not grow bigger; instead, they develop arthritis-like symptoms and various other pathologies. Perhaps better control of gene expression is required or even a more appropriate gene. The

Table 11.2 Efficiency of production of transgenic animals by microinjection of a growth hormone gene. (Adapted from Hammer *et al.* 1985.)

Animal species	No. of ova injected	No. of offspring	No. of transgenic offspring
Rabbit	1907	218	28
Sheep	1032	73	1
Pig	2035	192	20

results with human insulin expression in mice suggest that ultimately success will be achieved with animals of commercial value. If so, what genes might be manipulated? Certain histocompatibility genes might be good candidates for, if selected carefully, they could be used to enhance disease resistance. In sheep a good choice would be the Merino booroolla gene which confers increased ovulation rates and large litter sizes — up to seven lambs per litter — without detriment to body size or quality and quantity of wool. However, before these genes can be transferred the frequency with which microinjected eggs develop into healthy animals needs to be improved.

TRANSGENESIS IN MAN?

Ethics apart, are transgenic humans a possibility in the near future? The answer must be in the negative for three basic reasons. First, microinjection has a high failure rate even in experienced hands and this presupposes that sufficient human fertilized eggs could be obtained. Second, microinjection of eggs can produce deleterious effects because there is no control over where the injected DNA will integrate in the genome. A high level of spontaneous abortion may be tolerated in farm animals but it would not be acceptable in humans. Finally, there is the question of limited usefulness. Most of the serious genetic disorders result in death before puberty or infertility in homozygous patients. Therefore the worst case that can be imagined is that both parents are heterozygous and, on average, only one-quarter of the offspring will suffer the consequences of homozygosity. The question is, which ova are affected? Currently there is no way to tell and manipulating all ova poses far greater hazards!

GENE THERAPY IN HUMANS

As indicated earlier, only somatic cell gene therapy is being considered at present. This means that cells need to be extracted from the body, grown in culture, genetically manipulated and then reimplanted into the patient from whom the original tissue was taken. With current technology, this means that gene transfer in humans will be restricted to bone marrow and skin cells since only these cells can be maintained in culture. In turn, this severely restricts the number of genetic diseases for which therapy can be contemplated (Table 11.3). Studies with bone marrow cells are considerably more advanced than with skin cells and only the former are considered here.

Bone marrow consists of a heterogeneous population of cells, most of which are committed to differentiation into erythrocytes, lymphocytes, etc. (see Box). Less than 1% of nucleated bone marrow cells are the undifferentiated progenitor cells known as stem cells. In gene therapy it is the stem cells which would be the primary target for exogenous DNA. Since they are low in number and not easily recognizable, an efficient gene delivery system is needed. With bone marrow cells calcium phosphate-mediated DNA uptake is very ineffective: the transfection frequency is as low as 1 cell in 10^6-10^7. Approximately 10^{10} cells can be obtained from the marrow of a patient undergoing transplantation and only 10^7-10^8 of them will be stem cells. Thus, at best, only 100 stem cells would be transfected by the calcium phosphate method and these would have little effect when transferred back into the marrow of the patient unless they had some extraordinary growth advantage.

Retrovirus vectors (see p. 111) offer a number of advantages over calcium phosphate-mediated DNA uptake. First, up to 100% of recipient cells can be infected and express the cloned gene. Second, as many cells as desired can be infected. Third, under appropriate conditions the DNA can integrate as a single gene copy, albeit randomly. Finally, the infection with the retroviral vector does not appear to harm cells.

A MODEL SYSTEM FOR HUMAN GENE THERAPY

One candidate for human gene therapy is adeno-

Table 11.3 Target diseases for human gene therapy.

Protein	Disease state in absence of protein
Factor VIII	Haemophilia
Hypoxanthine-guanine phosphoribosyl transferase	Lesch–Nyhan disease
Pyrimidine nucleoside phosphorylase	Immunodeficiency
Adenosine deaminase	Combined immunodeficiency disease

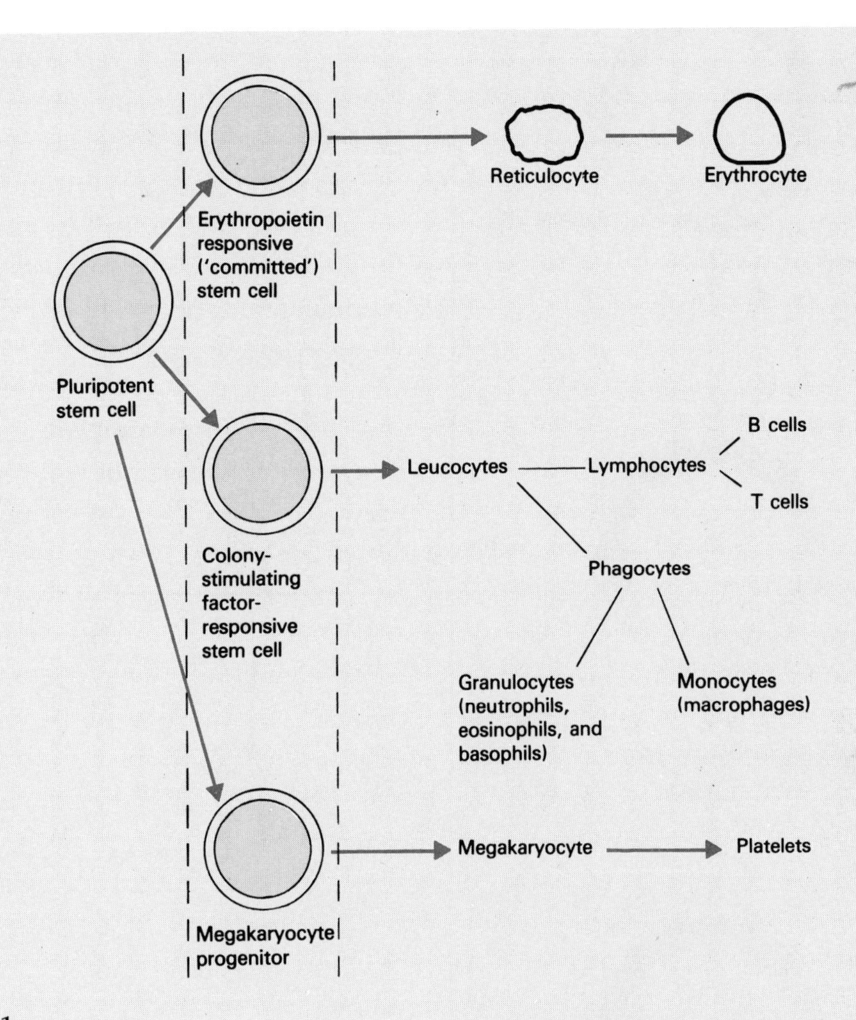

Diagram 1

The origin of the different blood cells

There are three formed elements in blood: erythrocytes (red blood cells), platelets, and leucocytes (white blood cells). These elements are formed in the bone marrow from pluripotent undifferentiated cells known as *stem* cells. In response to erythropoietin, a hormone produced by the kidney, a proportion of stem cells undergo maturation to produce the anucleate *reticulocyte*. Reticulocytes enter the circulation and mature into erythrocytes.

In response to a hormone, colony-stimulating factor, stem cells mature into five different kinds of leucocyte. Four of these different leucocytes are *phagocytic*, i.e. capable of digesting foreign debris in the bloodstream. The fifth class of leucocyte is the lymphocyte. There are two subsets of lymphocyte, the *B cells*, which produce antibodies, and *T cells*, responsible for the elimination of antibody-bound antigens.

Platelets are not cells but membrane-bound fragments of the cytoplasm of large cells known as megakaryocytes and which are found only in bone marrow.

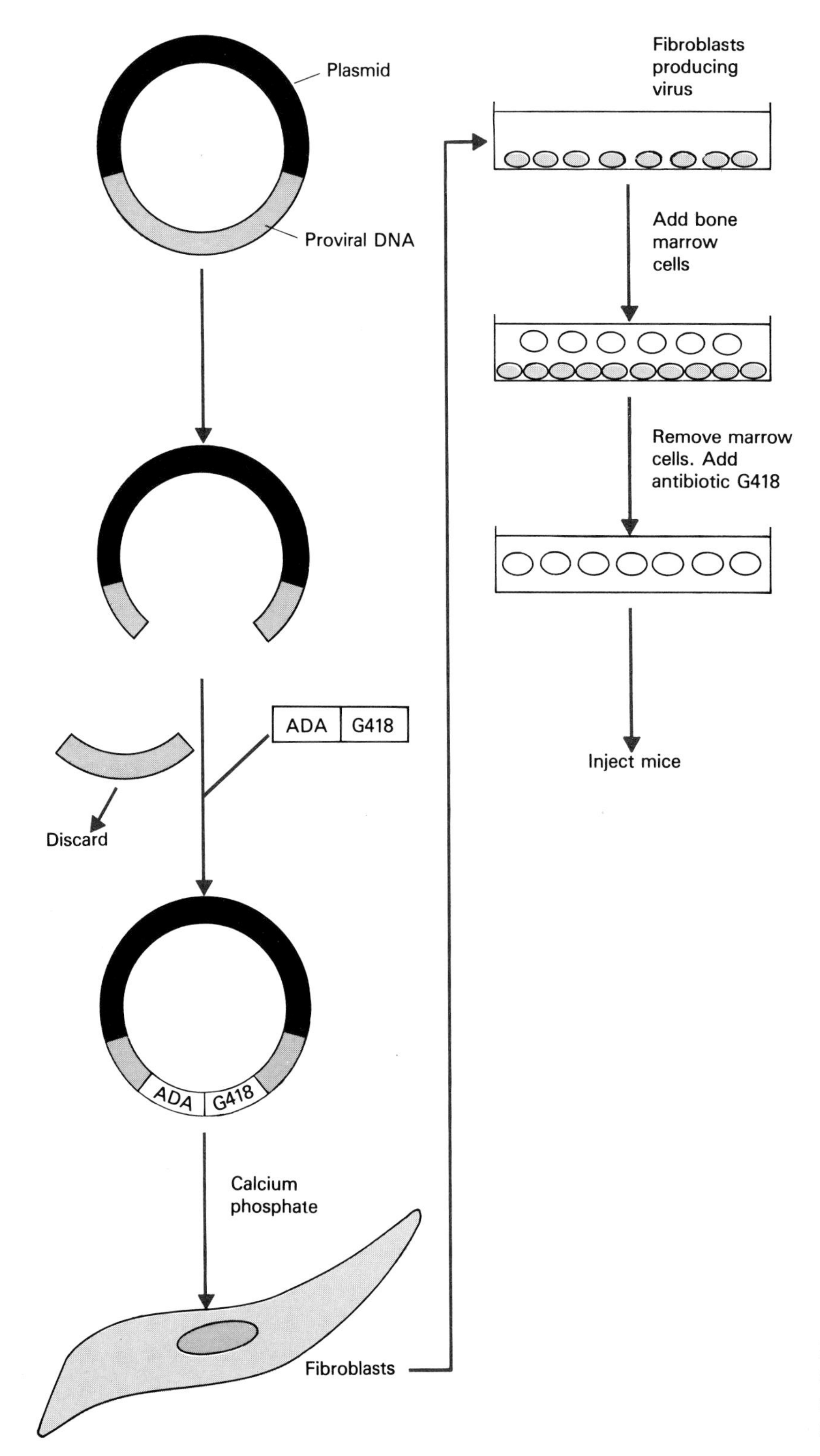

Fig. 11.4 Expression of human adenosine deaminase (ADA) gene in mouse bone marrow cells. See text for details.

sine deaminase (ADA) deficiency and as a model system the human adenosine deaminase gene has been cloned in bone marrow cells of mice. The protocol used is outlined in Fig. 11.4. Proviral DNA from a retrovirus was cloned in a plasmid and some of the proviral DNA replaced with the ADA gene and a gene encoding resistance to the antibiotic G418. The recombinant DNA was then introduced into fibroblasts in tissue culture using the calcium phosphate-mediated uptake method. Since the retroviral genome is now defective the fibroblast line selected was one in which the missing viral genes were incorporated into the cell genome. Once viral particles began to appear in the growth medium, mouse bone marrow cells were added. The marrow cells were subsequently removed and those carrying the antibiotic resistance gene selected by growth in the presence of G418. These marrow cells then were injected back into the mouse and synthesis of human ADA detected.

Despite this success, considerable improvements still need to be made in the vector systems. Retroviruses still are not reliable enough to be considered for human therapy. A major disadvantage of retroviruses is that they are prone to delete gene sequences. Another problem is that one retrovirus can exchange gene sequences with another retrovirus. Thus a retroviral vector might recombine with an endogenous viral sequence to produce infectious recombinant virus. The risk is small, particularly if murine retroviruses are used in humans for there is little homology between murine and primate retroviruses. Nevertheless, murine retroviruses are considered a hazard in pharmaceuticals produced using MABs (see p. 120)!

Further reading

GENERAL

Anderson W.F. (1984) Prospects for human gene therapy. *Science* **226**, 401–9.

SPECIFIC

Church R.B. (1987) Embryo manipulation and gene transfer in domestic animals. *Trends in Biotechnology* **5**, 13–19.

Clark A.J., Simons P., Wilmut I. & Lathe R. (1987) Pharmaceuticals from transgenic livestock. *Trends in Biotechnology* **5**, 20–4.

Dick J.E., Magli M.C., Phillips R.A. & Bernstein A. (1986) Genetic manipulation of hematopoietic stem cells with retrovirus vectors. *Trends in Genetics* **2**, 165–70.

Hammer R.E., Palmiter R.D. & Brinster R.L. (1984) Partial correction of murine hereditary growth disorder by germ-line incorporation of a new gene. *Nature* **311**, 65–7.

Hammer R.E., Pursel V.G., Rexroad C.E., Wall R.J., Bolt D.J., Ebert K.M., Palmiter R.D. & Brinster R.L. (1985) Production of transgenic rabbits, sheep and pigs by microinjection. *Nature* **315**, 680–3.

Selden R.F., Skoskiewicz M.J., Howie K.B., Russel P.S. & Goodman H.M. (1986) Regulation of human insulin gene expression in transgenic mice. *Nature* **321**, 525–8.

Steinmetz M. (1985) Immune response restored by gene therapy in mice. *Nature* **316**, 14–15.

Williams D.A., Lemischka I.R., Nathan D.G. & Mulligan R.C. (1984) Introduction of new genetic material into pluripotent haematopoietic stem cells of the mouse. *Nature* **310**, 476–80.

Part V
Plant Biotechnology

12/Plant Cell, Tissue and Organ Culture

The in-vitro culture of cells, tissues and organs of plants is one of the growth areas of biotechnology because of its potential to generate improved crops and ornamental plants. Such improvements have two components: the generation of genetic variation and the selection, maintenance and propagation of desired variants. The cultural techniques of plant cell, tissue and organ culture together with recombinant DNA technology are ideally suited to effecting both. At this point two popular misconceptions need to be corrected. First, these newer techniques will not appreciably shorten the time taken to introduce new plant varieties to the market place. This is because any new variety, however created, needs to be extensively field tested. Second, plant cell, tissue and organ culture are not synonomous; they are quite distinct techniques although they have many features in common. To appreciate each it is necessary to have a basic knowledge of the structure and development of flowering plants.

PLANT GROWTH AND DEVELOPMENT

When a seed germinates, a seedling emerges and grows into an adult plant with its many structures (Fig. 12.1). This plant consists of a stem and a root, each with an apical growth region, and many branches. In the stem the branches originate as buds which give rise to either leaves or flowers. As with stems and roots each bud also has its own apical growth region. Each apical growth region consists of *meristematic* cells which are undifferentiated cells that are the sources of all cell types in the primary tissues of a plant.

Plant development involves a succession of events necessary for formation of new tissues and these events involve different classes of regulation; for example leaves, flowers and fruit exhibit *determinate* growth, that is, growth ceases once they reach a certain size and shape. By contrast, roots and stems are *indeterminate* in growth for the meristems that produce them can continue to proliferate indefinitely provided conditions are suitable.

As the seedling emerges from the seed primary growth of the root occurs by multiplication of meristematic cells in the root apex. Above the meristem the cells grow in length without multiplication. Higher up, the outermost layer of elongated cells (the *epidermis*) can develop into root hairs whose function is the absorption of water and nutrients. As the root grows some of the root cells undergo differentiation into the phloem and the xylem. The *phloem* consists of cells with perforated ends whose function is the conduction of nutrient sap. The *xylem* consists of cells with reinforced walls whose function is water conduction. Between the phloem and the epidermis is a mass of undifferentiated cells known as the *cortex*. While the root undergoes its development so too does the shoot. Division of the meristematic cells of the shoot apex results in growth of the stem and some of these stem cells differentiate into developing leaves (*leaf primordia*). Between these primordia and the elongating stem are axillary buds, also with meristems, which will give rise to branches or flowers.

Division of the apical meristem produces vertical or *primary* growth; increased width or thickness comes from *secondary* growth. Secondary growth starts with the appearance between the phloem and the xylem of a layer of meristematic cells known as the *vascular cambium*. These cambium cells multiply laterally and differentiate into xylem and phloem. As the xylem increases in thickness the innermost part dies and becomes the *pith*. Between the cortex and the epidermis another layer of meristematic cells differentiates into the cork cambium whose growth produces the tough outer layer, which on a tree is known as *bark*.

PLANT TISSUE CULTURE

Tissue culture is the process whereby small pieces

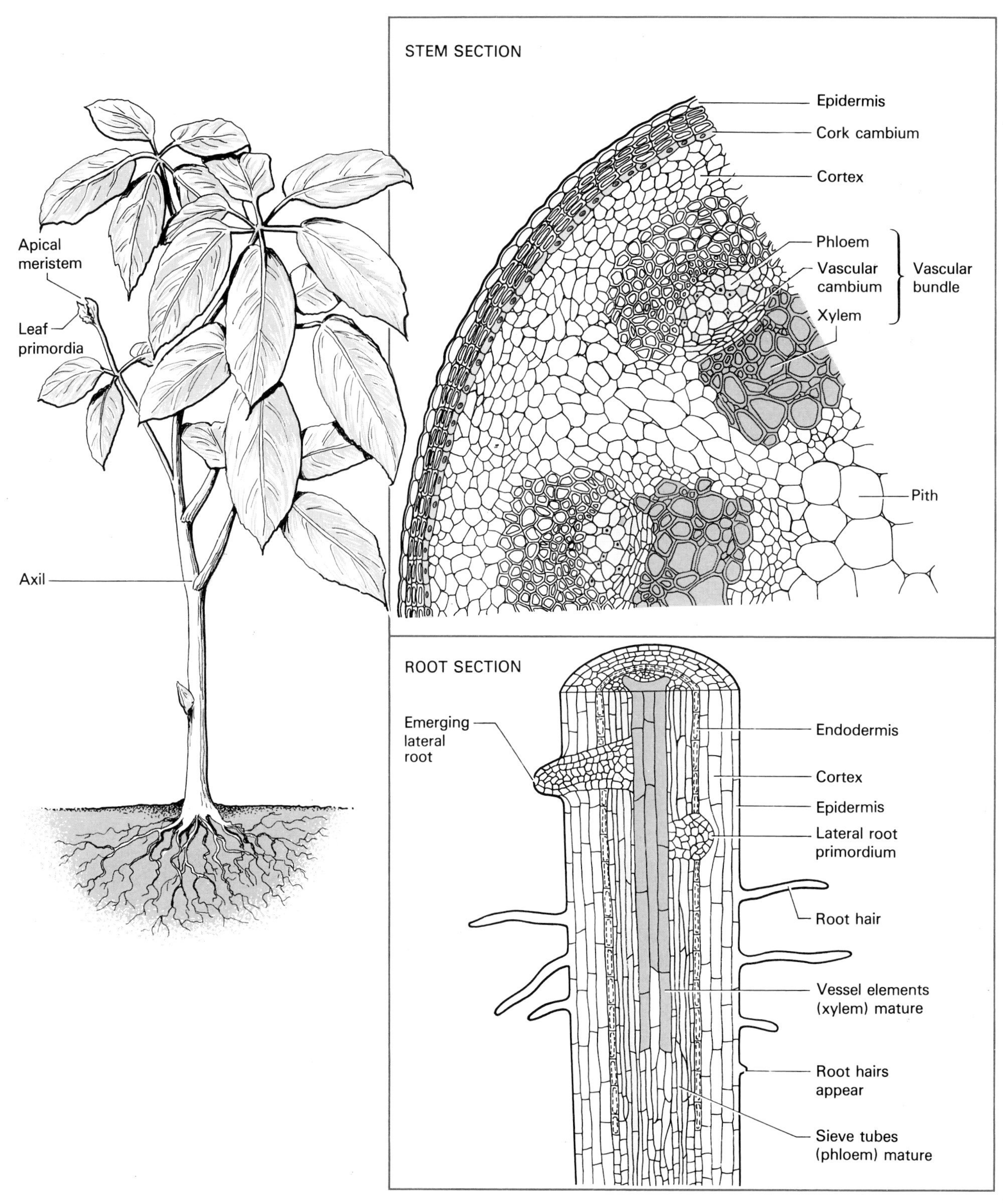

Fig. 12.1 The basic structure of a flowering plant. The insets show a horizontal and a vertical section through part of a plant to illustrate the structural detail.

of living tissue (*explants*) are isolated from an organism and grown aseptically for indefinite periods on a nutrient medium. For successful plant tissue culture it is best to start with an explant rich in undetermined cells, e.g. those from the cortex or meristem, because such cells are capable of rapid proliferation. The usual explants are buds, root tips, nodal segments or germinating seeds and these are placed on suitable culture media where they grow into an undifferentiated mass known as a *callus* (Fig. 12.2). Because the nutrient media used can support the growth of microorganisms the explant is first washed with a disinfectant such as sodium hypochlorite, hydrogen peroxide or mercuric chloride. Once established, the callus can be propagated indefinitely by subdivision (Fig. 12.3).

Fig. 12.2 Close-up view of a callus culture.

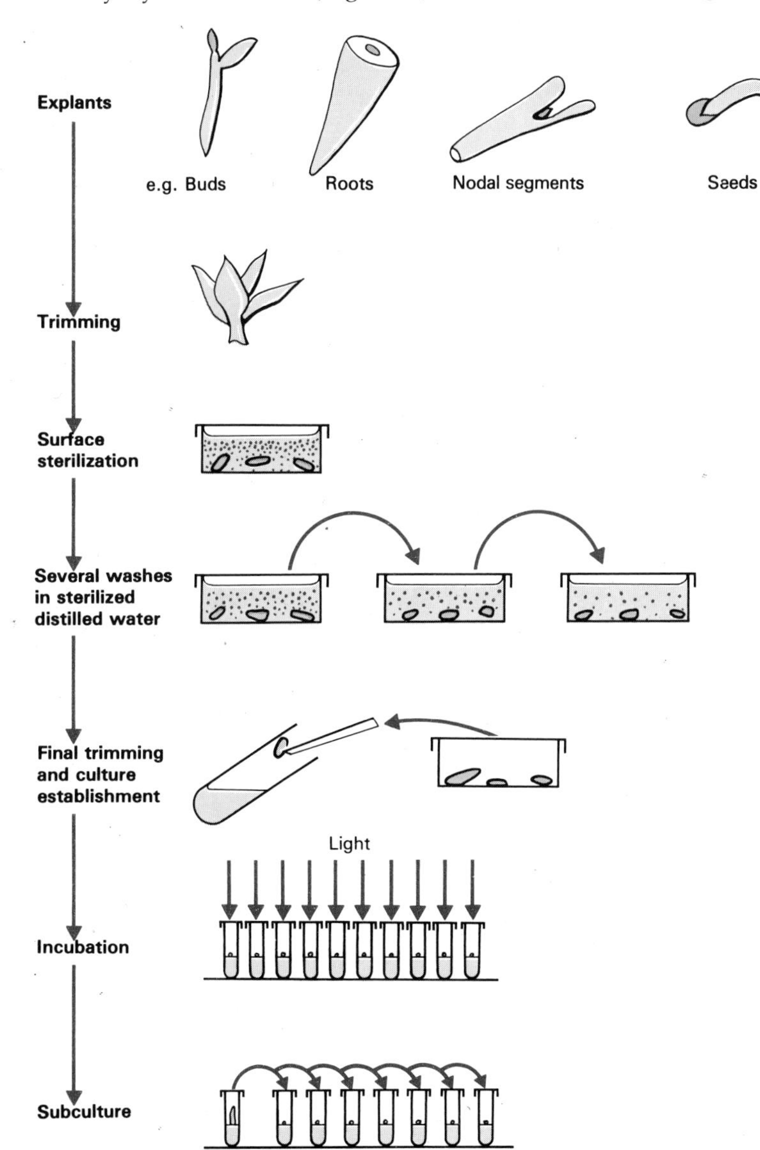

Fig. 12.3 Basic procedure for establishing and maintaining a culture of plant tissue. (After Mantell *et al.* 1985.)

For plant cells to develop into a callus it is essential that the nutrient medium contain plant hormones, i.e. an auxin, a cytokinin and a gibberellin (Fig. 12.4). The absolute amounts of these which are required vary for different tissue explants from different parts of the same plant and for the same explant from different genera of plants. Thus there is no 'ideal' medium. Most of the media in common use consist of inorganic salts, trace metals, vitamins, organic nitrogen sources (glycine), inositol, sucrose and growth regulators. Organic nutrients such as casein hydrolysate or yeast extract and a gelling agent are optional extras. The composition of a typical plant growth medium is shown in Table 12.1.

PLANT CELL CULTURE

When a callus is transferred to a liquid medium and agitated the cell mass breaks up to give a suspension of isolated cells, small clusters of cells and much larger aggregates (Fig. 12.5). Such suspensions can be maintained indefinitely by subculture but, by virtue of the presence of aggregates, are extremely heterogeneous. A high degree of genetic instability adds to this heterogeneity. Some plants such as *Nicotiana tabacum* (tobacco) and *Glycine max* (soybean) yield very friable calluses and cell lines obtained from these species are much more homogeneous and can be cultivated both batchwise and continuously. The cells can also be immobilized by the techniques described in Chapter 7.

When placed in a suitable medium isolated single cells from suspension cultures are capable of division. As with animal cells (see p. 102), for proliferation to occur conditioned medium may be necessary. Conditioned medium is prepared by culturing high densities of cells of the same or different species in fresh medium for a few days and then removing the cells by filter sterilization. Media conditioned in this way contain essential amino acids such as glutamine and serine as well as growth regulators like cytokinins. Provided conditioned medium is used, single cells can be plated out on solid media in exactly the same way as microorganisms; instead of forming a colony as do microbes, plant cells proliferate to give a callus. In this way it is possible to isolate and propagate mutant cell lines. The occurrence of such mutants (*somaclonal variation*) is proving to be of great benefit to the development of novel plants (see Chapter 13).

Protoplasts Protoplasts are cells minus their cell walls. They are very useful materials for plant cell manipulations because under certain conditions those from similar and contrasting cell types can be fused to yield somatic hybrids, a process known as *protoplast fusion*. Protoplasts can be produced from suspension cultures, callus tissue or intact tissues, e.g. leaves, by mechanical disruption or, preferably, by treatment with cellulolytic and pectinolytic enzymes. Pectinase is necessary to break up cell aggregates into individual cells and the cellulase to remove the cell wall proper. After enzyme treatment protoplast suspensions are collected by centrifugation, washed in medium without enzyme, and separated from intact cells and cell debris by flotation on a cushion of sucrose (Fig. 12.6). When plated onto nutrient medium protoplasts will in 5–10 days synthesize new cell walls and then initiate cell division.

PLANT ORGAN CULTURE

If apical shoot tips are surface sterilized and placed in a growth medium lacking plant hormones, they will develop into single seedling-like shoots. If instead the medium is supplemented with a cytokinin, axillary shoots will emerge from their normal positions in the leaf axils and produce a shoot cluster. Once such clusters have developed they can be subdivided into smaller clumps of shoots or separate shoots which will, in turn, form similar clusters when subcultured on fresh medium (Fig. 12.7). Provided the basic nutrient formulation is adequate for normal growth this subdivision process which is known as *micropropagation* can be undertaken every 4–8 weeks for an indefinite period. Indefinite culture of plant roots can be achieved in a similar manner but in this instance the nutrient medium should contain an auxin and no cytokinin.

REGENERATION OF PLANTS

Numerous species of plant can be maintained in cell or callus culture. Once in this state it is possible to isolate new genotypes by means of somaclonal variation, protoplast fusion or mutagenesis. However, such cell variants are of little value unless they can be induced to develop into intact plants. Cells which can differentiate into an intact plant are said

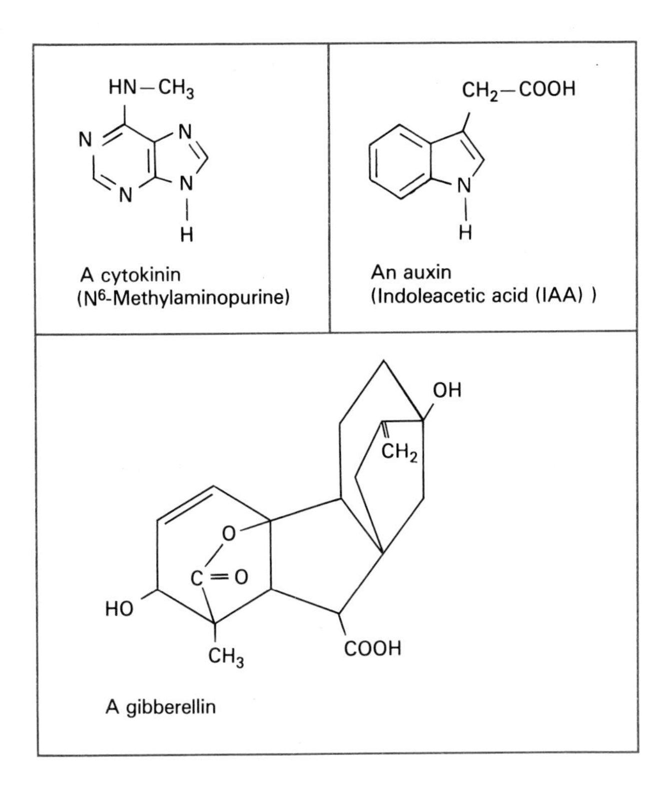

Fig. 12.4 The structures of some chemicals which are plant growth regulators.

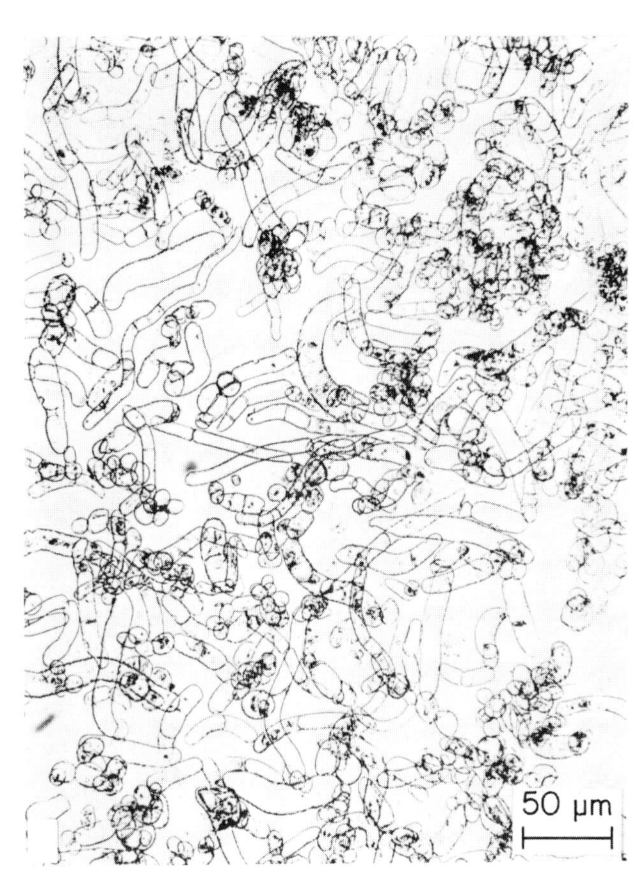

Fig. 12.5 Photomicrograph of plant cells in suspension culture.

Table 12.1 Composition of Murashige and Skoog (MS) culture medium.

Ingredient	Amount (mg/l)
Sucrose	30 000
$(NH_4)NO_3$	1650
KNO_3	1900
$CaCl_2.2H_2O$	440
$MgSO_4.7H_2O$	370
KH_2PO_4	170
$FeSO_4.7H_2O$	27.8
Na_2EDTA	37.3
$MnSO_4.4H_2O$	22.3
$ZnSO_4.7H_2O$	8.6
H_3BO_3	6.2
KI	0.83
$Na_2MoO_4.2H_2O$	0.25
$CoCl_2.6H_2O$	0.025
$CuSO_4.5H_2O$	0.025
Myo-inositol	100
Glycine	2.0
Kinetin (a cytokinin)	0.04–10.0
Indoleacetic acid	1.0–30.0

to be *totipotent* and totipotency is a property of undetermined cells. Undetermined cells are, by definition, capable of switching to different pathways of development depending on the environment in which they are located. As might be expected, undetermined cells are those cells which most readily develop into a callus.

As indicated earlier, the formation and maintenance of callus cultures requires the presence of a cytokinin and an auxin, whereas only a cytokinin is required for shoot culture and only an auxin for root culture. Therefore it is no surprise that increasing the level of cytokinin to a callus induces shoot formation and increasing the auxin level promotes root formation. Ultimately plantlets arise through development of adventitious roots on the shoot buds formed, or through development of shoot buds from tissues formed by proliferation at the base of rootlets (Fig. 12.8). The formation of roots and shoots on callus tissue is known as *organogenesis*. The cultural conditions required to achieve

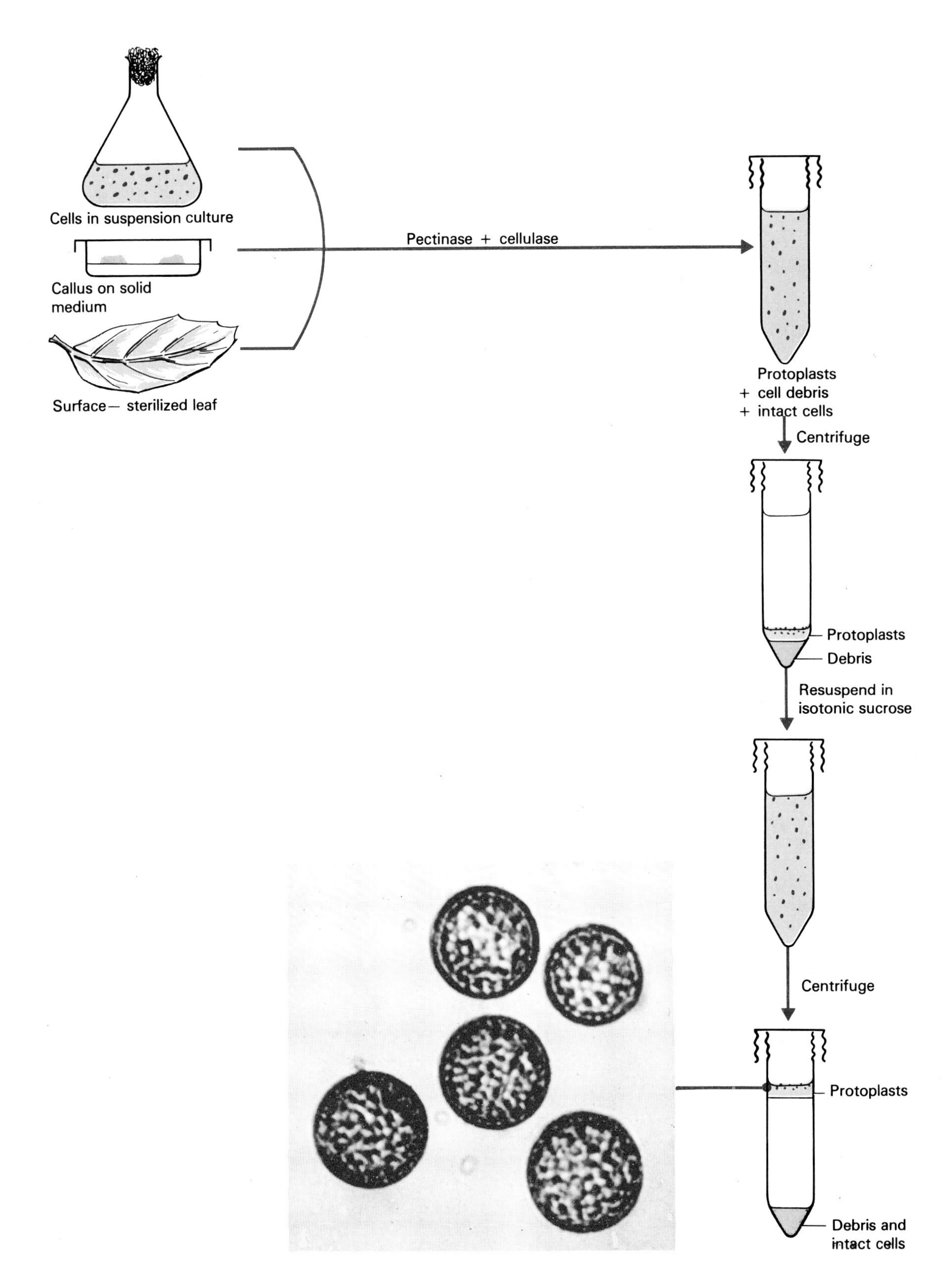

Fig. 12.6 Schematic outline of the enzymatic procedure used to isolate plant protoplasts. The inset shows a photomicrograph of protoplasts.

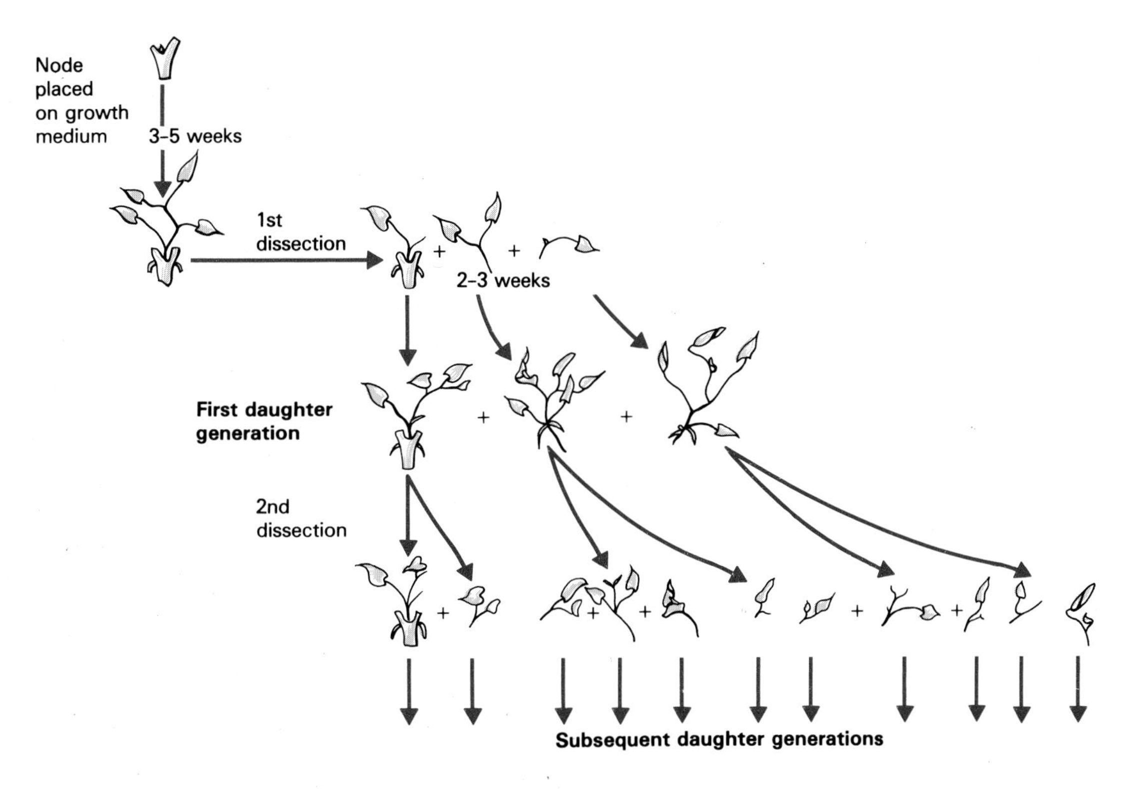

Fig. 12.7 Plant organ culture by means of axillary bud proliferation. (Reproduced courtesy of Dr S.H. Mantell.)

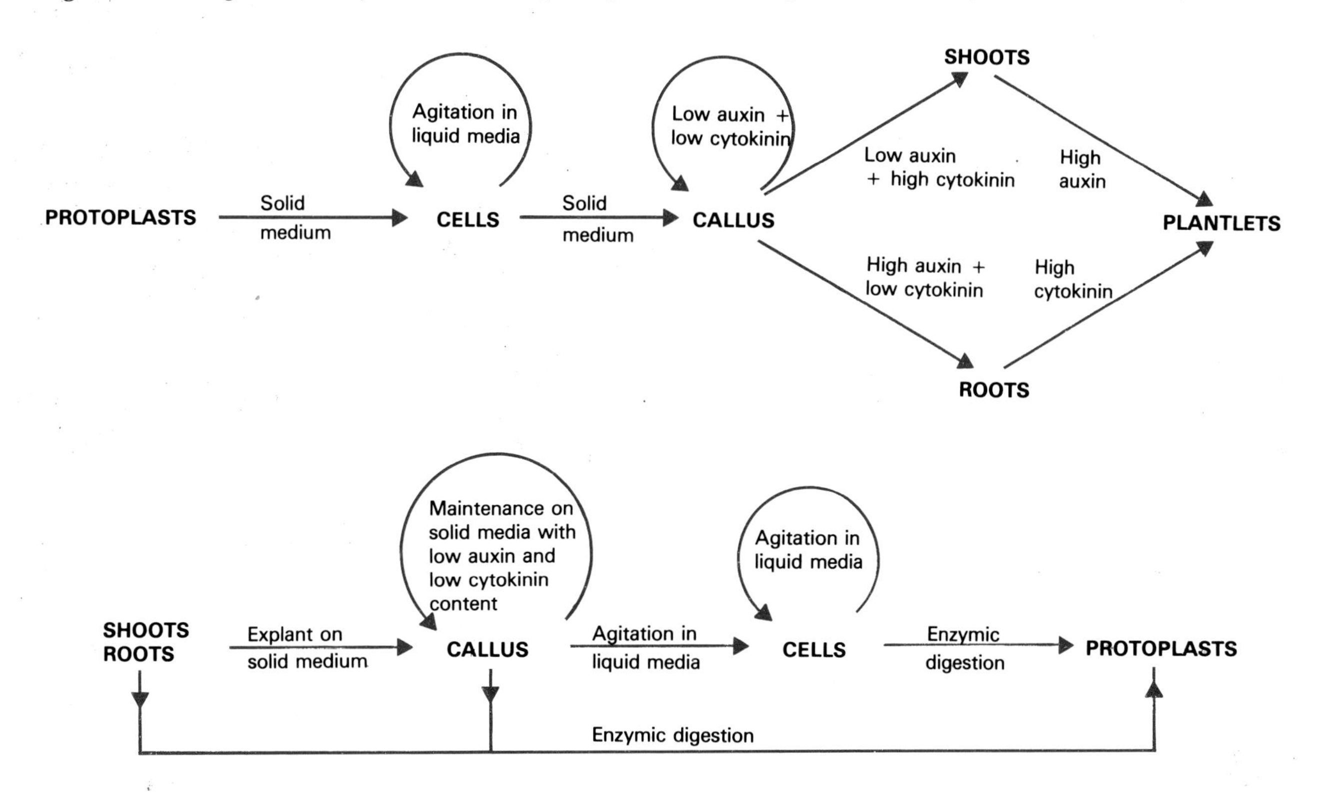

Fig. 12.8 Summary of the different cultural manipulations possible with plant cells, tissues and organs.

organogenesis vary from species to species, and have not been determined for every type of callus.

Under certain cultural conditions calluses can be induced to undergo a different development process known as somatic embryogenesis. In this process the callus cells undergo a pattern of differentiation, similar to that seen in zygotes after fertilization, to produce *embryoids*. Such cells are embryo-like but differ from normal embryos in being produced from *somatic* cells and not from the fusion of two germ cells. These embryoids can develop into fully functional plants without the need to induce root and shoot formation on artificial media. The embryogenic response leading to embryoid formation is stimulated when calluses which were established in medium in which 2,4-dichlorophenoxyacetic acid (2,4-D) was the auxin are transferred to 2,4-D-free medium containing reduced sources of nitrogen, e.g. ammonium salts.

ANTHER AND POLLEN CULTURE (see Box)

Pollen grains are determinate cells whose normal development involves the formation of a pollen tube and male gametes. When placed on suitable nutrient media, e.g. MS medium, most pollen grains follow this development pathway but a few grains will form a callus instead. Instead of culturing isolated pollen grains it is possible to culture the intact anthers containing the developing pollen and this results in the formation of embryoids directly from pollen grains. These embryoids can be induced to develop into whole plants which are true haploids (N.B. seeds of most flowering plants are diploids). As long as optimal conditions for donor plants and explants are provided, it is possible to obtain several hundred haploid plants *from a single anther*.

For successful anther and pollen culture it is essential to excise the flower buds at the correct time and usually this is at the time of the first mitotic division of the uninucleate microspore tetrads. Treatment of the excised buds or anthers prior to culturing on nutrient media can also increase the frequency at which pollen grains develop into plants and the treatment most often used is storage in the cold for several days to several weeks. Such cold stress probably raises the levels of growth regulators in the anthers. It should not be surprising, therefore, that inclusion of growth regulators in the culture medium can also be beneficial.

APPLICATIONS OF CELL, TISSUE AND ORGAN CULTURE

In this section only non-breeding applications of cell, tissue and organ culture will be discussed. The

Sexual reproduction in flowering plants (see Diagram 1)

The female reproductive structure is the pistil which consists of the ovary containing the ovule, a stalk-like style and the stigma which is the receptor for the pollen grain. The stamens are the male reproductive structures and consist of anthers (the pollen sacs) borne on filaments.

During male gametogenesis diploid microspore mother cells undergo meiosis to produce haploid microspores and these undergo mitosis and develop into pollen grains. Each pollen grain contains two nuclei; one, the tube nucleus, directs the formation of the pollen tube after pollination and then degenerates, the other, the generative nucleus, undergoes mitosis in the pollen tube to give two sperm nuclei.

During female gametogenesis, megaspore mother cells in the ovule undergo meiosis to produce four haploid megaspores. Three of these degenerate and the fourth undergoes three rounds of mitosis to produce a single cell, the embryo sac, containing eight nuclei. Two of these, the polar nuclei, after fusion with a sperm nucleus will generate the endosperm. Of the remaining six nuclei in the embryo sac, one will fuse with the other sperm nucleus to produce the diploid zygote. The remaining five nuclei in the embryo sac disintegrate. The zygote ultimately will develop into the embryo and, together with the endosperm, will make up the seed. The ovary develops into the fruit and contains the seeds.

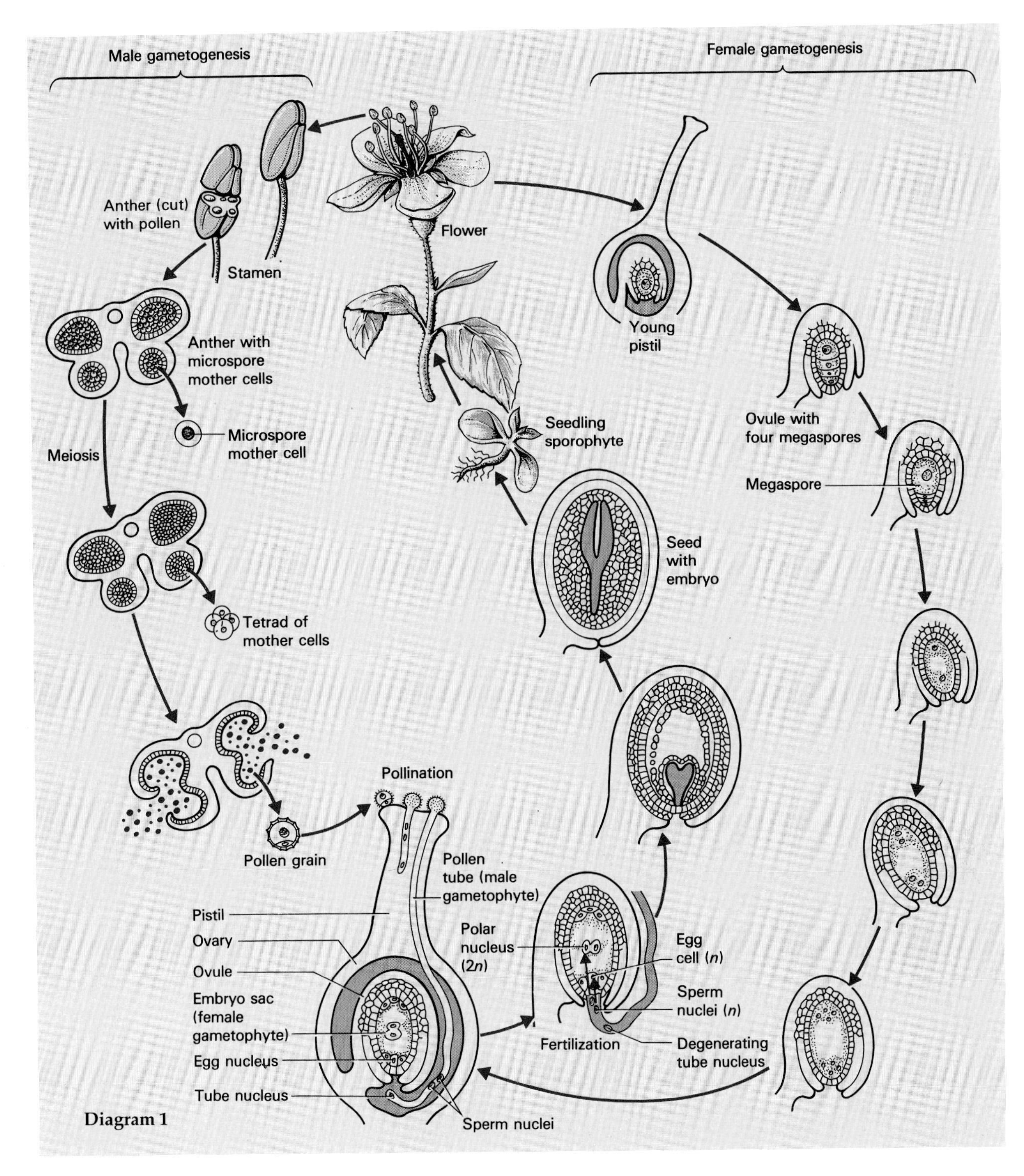

Diagram 1

applications in plant breeding are discussed in Chapter 13.

Production of fine chemicals For centuries plants have been an important source of precursors and products used in a variety of industries including those dealing with pharmaceuticals, food, cosmetics and agriculture (Table 12.2). Although in many cases natural plant products have been superseded by synthetic compounds, the plant kingdom is still a

Table 12.2 Some chemicals obtained from plants.

Product	Use	Source
Codeine	Analgesic	*Papaver somniferum*
Diosgenin, Sitosterol	Raw material for production of pharmacologically active steroids	*Dioscorea deltoidia*, *Zea mays*
Quinine	Antimalarial; embittering agent for food and drink	*Cinchona ledgeriana*
Digoxin	Treatment of cardiovascular disorders	*Digitalis* sp.
Scopolamine (hyoscine)	Treatment of nausea, especially motion sickness	*Datura stramonium*
Vincristine	Treatment of certain cancers	*Catharanthus roseus*
Atropine	Treatment of cardiac arrhythmias. Dilation of the pupils of the eye	*Atropa belladonna*
Reserpine	Treatment of hypertension	*Rauwolfia serpentina*
Pyrethrin	Insecticide	*Chrysanthemum* sp.
Jasmine	Perfumery	*Jasminum* sp.
Saffron	Food colourant and flavouring agent	*Crocus sativus*
Menthol	Flavouring	*Mentha piperita*

major contributor of speciality chemicals with a market value measured in billions of dollars. It is worth remembering that the more structurally complex the compound, and in particular the number of stereochemical centres in the molecule, the less likely it is that plant production will be replaced by chemical synthesis.

The demand for plant-derived chemicals is largely from the developed countries but they are produced in third-world countries where political instability and lack of funds means that social disorder or plant disease could interrupt supplies at any time. Could plant cell culture provide an alternative source of such compounds? Given current technology an average process would have a yield of 20 g dry weight of cells per litre and a productivity of 10 g dry weight per litre per day. If 1% of the cell mass were product then, for the process to be economically viable, the product would need to have a value of at least $500/kg. Nor is product value the only criterion; the size of the market is equally important. Not surprisingly, only one commercial process has emerged, the synthesis of shikonin by cells of *Lithospermum erythrorhizon*. Part of the problem is that many of those plant products which are of high value are not one but a mixture of compounds. Oils used as plant fragrances are a good example. Although all the various ingredients may be produced by culturing cells of the appropriate plant species, the proportion of the different components relative to one another will be wrong.

Where a plant product is a single chemical species then improving the yield per unit cell mass would greatly improve the economics of cell culture. In this regard it should be noted that the yield of shikonin, the only plant cell culture product commercially available, is 14% of cell dry weight. One way of increasing product yield might be to mimic the methods used to increase penicillin production (see p.74) and isolate high-yielding plant cell lines. Before genetic engineering techniques will be of much value it will be necessary to have a considerable amount of information about the biochemistry and genetics of product formation.

Over 1500 new compounds are identified in plants every year and it has been suggested that some of these could be developed into products produced in cell culture. The problem with this approach is that it is technology driven and not market led. By definition the compounds will be of high value and the most likely customer is the pharmaceutical industry. However, in recent years this industry has directed its drug-seeking activities at designing chemicals specifically to interact with key enzymes or biological effectors. The use of natural compounds is not fashionable.

Production of pathogen-free plants Crop and ornamental plants can be infected by a wide variety of microbial pests: viroids, viruses, mycoplasmas, bacteria, fungi and nematodes. Such infections greatly reduce the yield, vigour and quality of plants. For example, virus infection of ornamental plants reduces the size and number of blooms produced and infection of fruit crops reduces yield by up to 90%. In some cases infection can be eliminated by micropropagation of unaffected parts of the plants and this approach is particularly feasible if the problem is caused by bacteria, fungi or nematodes. Viroid, virus and mycoplasma infections are much more troublesome since they tend to be systemic. More specialized methods are required for their elimination and these methods are based on plant regeneration.

The basic method of obtaining virus-free plants is culture of apical meristems. If a small enough piece is taken and cultured, preferably less than the first half millimetre of the growing tip, it often is possible to obtain virus-free material which can be proliferated *in vitro* and then used to regenerate virus-free plants. An alternative method is to take larger explants of apical meristem and subject them to heat treatment to reduce or eliminate virus replication. Suitable treatments are 30–37 °C for 10–14 days or 50–60 °C for 5–10 min. The third method is to incubate apical meristem explants on suitable culture media containing malachite green or thiouracil which can reduce virus replication.

LARGE-SCALE PLANT PROPAGATION

If a plant can be propagated vegetatively, then it is a relatively easy matter to increase the number of plants available for cultivation. However, if a plant can only be propagated from seed, then increasing plant numbers can be a lengthy process. This is particularly true if one is dealing with shrubs and trees which may not flower until several years after seed germination. In such cases the use of in-vitro methods of propagation can be invaluable. In practice, an explant may be taken, induced to form a callus on solid media and then a suspension culture derived by inoculating the callus into a liquid medium (Fig. 12.9). After the cells have increased in number they are plated out and somatic embryoids

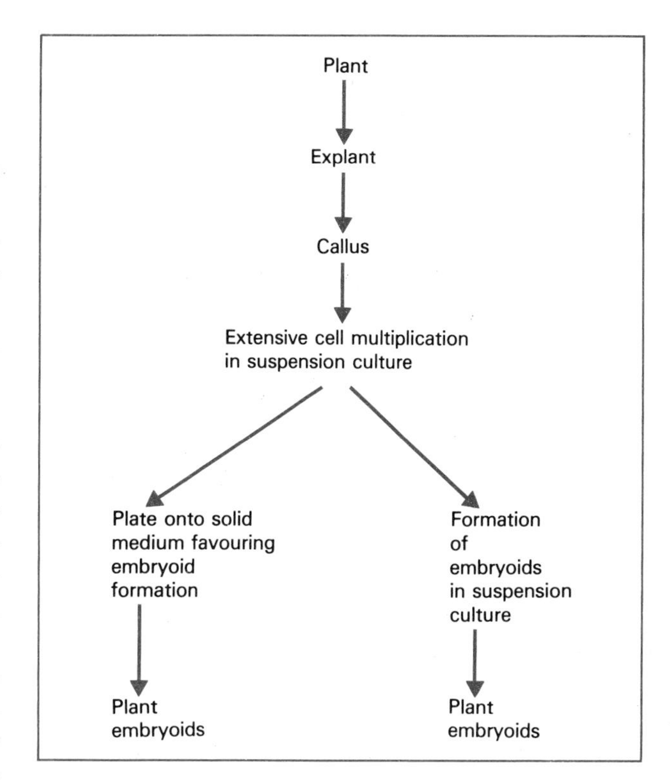

Fig. 12.9 Schematic outline of the method used for mass propagation of plants using in-vitro culture techniques.

selected. On cultivation these embryoids will develop into regular plants. Large-scale clonal propagation of this kind is being used commercially with a number of tropical plantation crops, e.g. oil palms.

Another micropropagation method which is being used commercially is the production of minitubers. When axillary shoot cultures of potato are cultivated in the presence of appropriate levels of cytokinins and gibberellic acid they form large numbers of very small tubers (*minitubers*). These minitubers can be sown directly in the field and will generate normal plants. The advantages of this method of micropropagation can be seen from the data in Table 12.3.

PLANT GERMPLASM BANKS

In this century plant breeders have made enormous contributions to increased food production throughout the world. One important part of their work is the introduction of genetic diversity by intercrossing or mating selected germplasm with outstanding characters that complement one another. One source of this genetic diversity is from related species

Table 12.3 Comparison of traditional and tissue culture methods for propagating tubers. The reduced number of plants obtained from the minitubers is an allowance for 5% loss on establishment in soil. (Reproduced from Mantell *et al.* 1985.)

Traditional	Micropropagation
Year 1	
100 g tuber	100 g tuber
↓	↓
One mature plant	One preconditioned plant
│	↓
│	10 nodal cultures
↓	↓
1600 g tubers	Shoot multiplication
↓	↓ *Dormant*
----------------------------	650 000 minitubers --------------- *season*
Year 2	
16 mature plants	617 500 plants
↓	↓
16 × 600 g tubers = 25.6 kg	617 500 × 500 g tubers = 308 750 kg

occurring in the wild, frequently in the less well-developed countries of the world. However, the numbers of wild species and their natural habitats are disappearing rapidly in these countries as a result of growing urbanization and an increase in the amount of land being cultivated. There is a general fear that potentially valuable germplasm is being lost irretrievably. As a consequence there is a growing use of gene banks. For many crops these comprise seed stores in which representative stocks of seed are held at a reduced temperature and humidity. The stocks are tested periodically for loss of viability and, if necessary, replenished by re-growth and harvesting.

There are two problems associated with seed banking. First, it cannot be used with so-called *recalcitrant* seeds. Such seeds are large, succulent, lack a dormancy phase and are incapable of surviving low temperatures and/or dehydration. Until now they have been maintained in small plantations set aside as *field gene banks*. Second, seed storage is not applicable for plants which are propagated vegetatively. Vegetative propagules such as bulbs and tubers can be stored but not for long periods of time.

As a result of recent advances, cell, tissue and organ culture is being considered as a means of germplasm storage. Two approaches are being adopted. One is *slow growth* involving the depression of metabolism by physical or chemical means and the other is the suppression of growth by storage at low temperatures (*cryopreservation*). Slow growth is only applicable to organized shoot cultures, whereas cryopreservation has proved most successful for cell cultures and shoot tips. Whatever method is used, it is essential that genetic variability should not occur during regeneration of the intact plant and this cannot be guaranteed.

Further reading

GENERAL

Mantell S.H., Matthews J.A. & McKee R.A. (1985) *Principles of Plant Biotechnology*. Blackwell Scientific Publications, Oxford.

Sen S.K. & Giles K.L. (1982) *Plant Cell Culture in Crop Improvement*. Plenum Press, New York.

Yeoman M.M. (1986) *Plant Cell Culture Technology*. Blackwell Scientific Publications, Oxford.

SPECIFIC

Curtin M.E. (1983) Harvesting profitable products from plant tissue culture. *Bio/technology* **1**, 649–57.

Farnum P., Timmis R. & Kulp J.L. (1983) Biotechnology of forest yield. *Science* **219**, 694–702.

Giles K.L. & Morgan W.M. (1987) Industrial scale plant micropropagation. *Trends in Biotechnology* **5**, 35–9.

Klausner A. (1985) Researchers cotton on to new fiber findings. *Bio/technology* **3**, 1049–51.

13/Biotechnology and Plant Breeding

Conventional methods of plant breeding have been remarkably successful in generating improved varieties of many of the major world crops. Most of the improvement has been in yield and has occurred by the breeding and selection of lines resistant to pests and diseases, which can tolerate stressful environments and which are larger or are heavier croppers. There have also been improvements in the intrinsic quality of the crop, e.g. taste, protein content, etc. The long-range targets of biotechnology as applied to plant breeding are exactly the same; that is, the creation of new plant varieties with increased vigour and yield, the incorporation of value-added traits, and the development of disease-, herbicide- and pesticide-resistant crops.

Scientifically the most exciting developments in plant breeding are centred on the use of recombinant DNA to introduce new characteristics into plants and it is these developments which form the basis of part of this chapter. However there are a number of non-recombinant approaches which are just as useful and which are already bringing new crops to the marketplace.

Biotechnology and plant breeding: Non-recombinant approaches

Basically, there are two non-recombinant approaches which merit discussion. The first of these is the exploitation of spontaneous or induced variation in cultured plant cells or tissues. The second is the use of intra- and inter-specific protoplast fusion to mediate genetic exchange.

SOMACLONAL VARIATION

Somaclonal variation technology takes advantage of the naturally occurring genetic variation that appears in plants regenerated from somatic cells grown in tissue culture. This variation can pre-exist in the explant tissue but more usually it arises during the tissue culture procedure itself. The genetic events which occur include changes in nucleotide sequence, chromosome structure and chromosome number and the phenotypic variability which results can also occur in the regenerated plant and be heritable. However some of the phenotypic variability can result from the physiological response of the cells to the environment of the culture dish or vessel as well as from *epigenetic* changes. Epigenetic events reflect physiologically altered levels of gene expression that are relatively stable in that they persist through mitosis to be expressed by daughter cells. In contrast to altered phenotypes having a genetic basis, those resulting from epigenetic changes tend not to be expressed in regenerated plants or their progeny. Operationally, genetic changes are distinguished from epigenetic changes by their ability to be transmitted following sexual crossing.

The variability present in cell cultures is ultimately visible in populations of regenerated plants. Simple visual examination can be used to identify gross morphological changes such as plant height, growth habit, leaf shape and size, flower morphology and pigmentation. The frequency of visible variability is sufficiently high such that most individuals are altered in some way but not so high that the majority of those individuals possess deleterious alterations. However, most of the desired phenotypic traits are not associated with morphological changes and screening procedures have to be used. For example, early blight-resistant clones of a particular potato cultivar could only be identified by inoculating leaves of regenerated plants with toxin derived from *Alternaria solani*, the early blight fungus. In addition, where a crop is usually seed-propagated (e.g. cereals such as wheat, oats) rather than vegetatively propagated (e.g. potatoes) it is essential to show that the genetic change giving the desired trait has not caused infertility. Such infertility can arise as a result of chromosomal rearrangements and can only be detected by attempting sexual crosses.

The genetic diversity of plants emerging from disorganized callus tissues provides the breeder with a means of introducing variability into established cultivars without the use of sexual crosses (Table 13.1). However, conventional screening is still required and with it the need for large amounts of land and labour. What would be particularly advantageous is a means of directly selecting the desired phenotypes. So far there has been some limited success in this area by the application of some of the methods developed for selecting mutants of microorganisms. Thus tobacco cells resistant to methionine sulphoximine and maize cells resistant to *Dreschslera maydis* toxin have been selected *in vitro* and plants regenerated from them have been resistant to wildfire disease and southern leaf corn blight, respectively. Mutagenesis can be used to increase the frequency with which such mutants occur. However, many traits of agronomic importance such as grain quality, pest resistance, etc., cannot be selected directly. Not only are these traits not expressed by cultured cells but the absence of an understanding of the molecular and cellular bases of these traits precludes the development of in-vitro selection schemes. In the immediate future in-vitro methods may be beneficial for the selection of plant mutants with improved amino acid content. As with microbial cells (see p. 71) this can be achieved by deregulating the biosynthetic pathways and some examples are shown in Table 13.2.

Table 13.1 Some useful plant variants obtained through somaclonal variation.

Plant	Improvement
Carrots	Improved 'snacking' characteristics (sweetness, crunchiness, crispness)
Celery	Improved 'snacking' characteristics
Potato	Resistance to late blight (*Phytophthora infestans*) and early blight (*Alternaria solani*)
Sugar cane	Resistance to eyespot disease (*Helminthosporium sacchari*), downy mildew and Fiji disease (leafhopper-transmitted virus). Increased sucrose content
Tomato	Increased solids content

Table 13.2 Some useful plant mutants obtained following selection of mutant cells *in vitro*.

Plant	Mutation	Selection procedure
Maize	Resistance to southern corn leaf blight	Resistance to toxin produced by the causative agent (*Dreschslera maydis*)
Maize	Increased levels (×100) of threonine in amino acid pool of maize kernels. Total threonine content increased by 50%	Resistance to growth inhibition by a mixture of lysine and threonine
Tobacco	Resistance to wildfire disease (*Pseudomonas tabaci*)	Resistance to methionine sulphoximine which is an analogue of the toxin produced by causative organism
Tobacco	Salt tolerance	Resistance to high levels of sodium chloride
Tobacco	Herbicide resistance	Resistance to herbicide incorporated in culture media

PROTOPLAST FUSION

Often it is desirable to cross two related species that are sexually incompatible. Sexual crosses as used in conventional plant breeding clearly are impossible but the same goal can be achieved by fusing the relevant plant protoplasts.

Plant protoplasts isolated according to the method described in Chapter 12 (see p. 142) can be induced to fuse with each other even when derived from different species. Two basic methods are used. Certain chemicals such as polyethylene glycol, dextran and poly-L-ornithine act as *fusogens* in that they promote protoplast fusion. Alternatively *electrofusion* can be used. In this procedure protoplast adhesion occurs in a non-uniform electrical field and fusion occurs when a short pulse of direct current is applied. When the two protoplasts first fuse a cell is produced which contains both nuclei and both cytoplasms. If the parent cells were not identical, then the product is a *heterokaryon*. If the two nuclei subsequently fuse, a mononuclear cell known as a *synkaryon* is produced. Although fusion also brings the plastids (chloroplasts and mitochondria) together, they do not stay together and eventually the plastids from only one plant predominate (Fig. 13.1).

After fusion the nuclear and cytoplasmic genomes reassort and recombine resulting in a wide array of gene combinations not attainable through conventional breeding. The ability to transfer plastids is especially promising since traits such as cytoplasmic male sterility (see Box, p. 152), disease resistance and herbicide resistance are encoded by the organelles and not the nuclear genome. The biggest disadvantage is that the scientist has little control over what genetic information is retained and what is eliminated. When fusions are made between protoplasts from totally unrelated species, e.g. tobacco and soy-

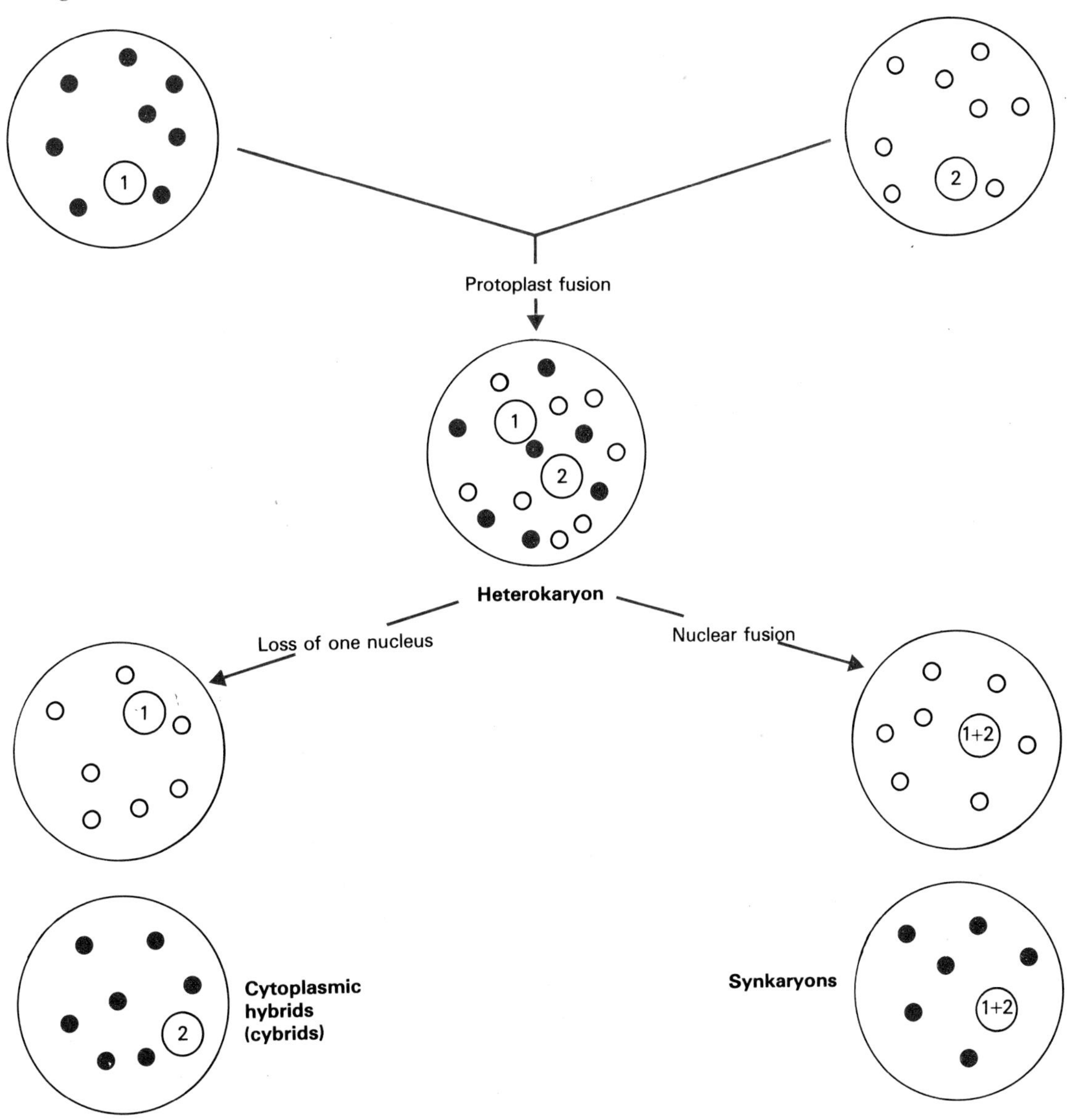

Fig. 13.1 The different kinds of hybrids which can be obtained following fusion of two different protoplasts. The nuclei of the two parental cells are labelled 1 and 2 and the plastids are shown as either red or black circles.

Cytoplasmic male sterility

Many important crops such as maize, barley, wheat, sugar cane, sugar beet and onions are grown commercially from hybrid seeds. The importance of outbreeding to form hybrids, rather than inbreeding by self-fertilization, is that the hybrid progeny usually grow more vigorously and give higher yields of foodstuff. The production of hybrid seeds requires carefully controlled pollination. Maize, for example, is self-fertile and it is essential to block self-pollination if hybrid seed lines are to be maintained. Originally labour was employed to remove the tassels (pollen-producing part) of maize plants by hand. Today the cost of this would be prohibitive and male-sterile plants are used instead.

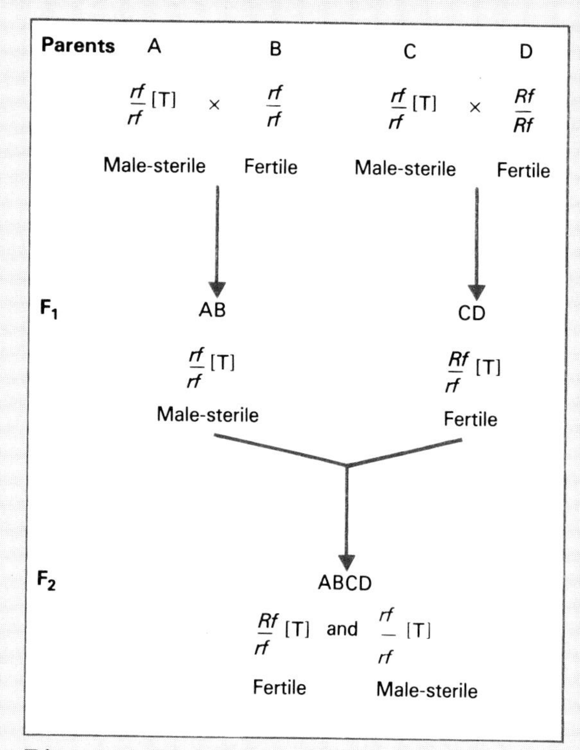

Diagram 1 The use of male-sterile plants in the production of hybrid maize. Note that only plants produced by outbreeding are shown. Plants can also be produced by inbreeding, i.e. B with B, D with D and CD with CD. However, inbred plants can be distinguished from outbred plants and the cobs not collected from them.

The maize seed which is sold commercially is the product of two sexual crosses involving four inbred lines represented in the diagram as A, B, C and D. Plants A and C carry a factor T (for Texas where it was discovered) which confers pollen sterility and these plants must be fertilized with pollen from another plant line, e.g. B or D. Pollen fertility can be restored by a dominant nuclear gene *Rf*. The F_1 hybrids both carry the factor T but whereas the AB hybrids are male-sterile, the CD hybrids produce fertile pollen because they carry the gene *Rf*. In the F_2 generation the AB hybrid can only be pollinated by the CD hybrid although CD can fertilize itself. By careful choice of genetic markers it is possible to distinguish between the AB hybrids and the CD hybrids and only select corn cobs (source of seed) from the AB plants. Half of the seed in these cobs will be male-sterile and half will be fertile. When the seeds are sown there will be ample pollen from the fertile plants to pollinate the male-sterile plants.

The cytoplasm of plant cells contains chloroplasts and mitochondria and these organelles possess DNA and their own protein-synthesizing machinery. There is now good evidence that the mitochondrial genome is involved in the phenomenon of male sterility; for example, restriction endonuclease analysis has shown differences between mitochondrial DNA from normal and male-sterile plants. Also, mitochondria from plants carrying the T factor synthesize a polypeptide of molecular weight 13 000 not found in normal plants. In the presence of the restorer gene, *Rf*, this polypeptide is not synthesized.

One disadvantage of male-sterile lines of corn is that they are particularly susceptible to southern corn leaf blight.

bean, the nuclei either fail to fuse or else one set of chromosomes is lost following fusion. In contrast, fusions between sexually incompatible members of the same family have produced hybrid plants that retain some chromosomes from both parents (Table 13.3). The best known of these is the 'pomato' which is a potato–tomato somatic hybrid. Somatic hybrid plants such as the pomato are not of imme-

Table 13.3 Examples of somatic hybrids created by interspecific protoplast fusion.

Lycopersicon esculentum (tomato)	× *Solanum tuberosum* (potato)
Datura innoxia (thorn apple)	× *Atropa belladonna* (deadly nightshade)
Arabidopsis thaliana (thale cress)	× *Brassica campestris* (turnip)
Petunia parodii (petunia)	× *Petunia parviflora* (petunia)

diate value. Not only do many combinations prove infertile or fail to set seed but they need to be extensively back-crossed to eliminate unwanted portions of the alien genome. What they do provide is a means for introducing genes from unconventional sources.

ANCILLARY TECHNIQUES

Incompatibility, which is the failure of pollen to fertilize an ovule, is a major obstacle in conventional plant breeding programmes. There are a number of causes of incompatibility but common ones are failure of the pollen tube to develop and abortion of the young embryo. One solution to this problem is to fertilize the ovule *in vitro*. Pistils are cultured on nutrient media and, after removing part of the ovary wall, fertilization is effected by placing the pollen directly on the exposed ovule. Another cause of incompatibility is failure of the embryo to develop because nutrient starvation prevents the endosperm from developing. In this instance plants can be recovered for breeding purposes by culturing the embryo on nutrient media. An essential requirement in many plant breeding programmes is plant stocks homozygous for genes of interest. Such stocks breed true but are laborious to prepare. Consider the cross AA BB cc dd × aa bb CC DD. All the progeny will have the genotype Aa Bb Cc Dd, i.e. will be heterozygous for all four loci. In order to obtain a pure-breeding stock of genotype AA BB CC DD up to five generations of back-crossing may be necessary. A much simpler alternative is to use anther or pollen culture. As has been discussed (see p. 144), plants regenerated from anther and pollen cultures are haploid but it is a simple matter to double the chromosome complement with colchicine treatment. Colchicine interferes with mitosis and effects chromosome doubling by preventing sister chromatids separating at anaphase.

CURRENT STATUS OF THE APPLICATION OF CELL, TISSUE AND ORGAN CULTURE TO PLANT BREEDING

The plant breeding industry has not been slow to recognize the advantages of these non-recombinant approaches. Specific examples include breeding disease-resistance in sugar cane, haploidization in potato, rice and wheat breeding, somaclonal variation in potato, rice, wheat and lucerne (alfalfa), in-vitro selection of herbicide-resistant cells of tobacco and somatic hybridization to develop acceptance of cytoplasmic male sterility in crops such as rape.

RECOMBINANT APPROACHES TO PLANT BREEDING

Given the success of conventional plant breeding and the potential benefits of such biotechnological approaches as somaclonal variation and protoplast fusion why bother to use recombinant methods? Two answers can be given to this question. First, genetic engineering can be used to introduce genes into a plant which do not exist in any member of the same plant family, or even in any plant. Second, gene manipulation in the guise of protein engineering could be used to make substantial alterations to key cellular proteins, changes which are unlikely to occur spontaneously.

Cloning vectors are required for genetic engineering of plants and two different kinds are available: those based on plant viruses and those derived from the Ti plasmid of the plant pathogen *Agrobacterium tumefaciens*. In addition, there must be a means of introducing the recombinant DNA into plant cells. Plant protoplasts can be induced to take up naked nucleic acid and the methods used are essentially no different from those used with animal cells (see Table 11.1, p. 127). However, handling protoplasts is technically tricky and a much easier method is available with the Ti plasmid (see p. 155).

If genetically engineered plants are to be used

commercially, then the following requirements must be satisfied:

1 introduction of the gene(s) of interest to all plant cells;

2 stable maintenance of the new genetic information;

3 transmission of the new gene to subsequent generations;

4 expression of the cloned genes in the correct cells at the correct time.

The extent to which these conditions are met by the two different vector systems is outlined below.

PLANT VIRUSES AS VECTORS

At first sight plant viruses appear very attractive as cloning vectors because not only can they infect intact plants but nucleic acid purified from them can also be infectious. However, not just any plant virus can be used for there are a number of other criteria which need to be satisfied. First, the virus must be able to spread from cell to cell via the plasmodesmata. Second, introduction of foreign nucleic acid into the virus genome might make it too large to be packaged in viral particles; thus the viral nucleic acid must be able to replicate and spread from cell to cell in the absence of viral coat protein. Third, the modified viral genome should elicit little or no disease symptoms in infected plants. Fourth, it would be preferable if the virus had a broad host range. Finally, since most genetic engineering centres around the in-vitro manipulation of DNA it is preferable that the virus vector has a DNA genome. Only two groups of plant viruses, the Caulimoviruses and the Geminiviruses, have DNA genomes.

Caulimoviruses as vectors Caulimoviruses are a group of spherical viruses (Fig. 13.2) which contain a circular double-stranded DNA genome. The archetypal member is Cauliflower Mosaic Virus (CaMV) hence the name of the group. The Caulimoviruses are widely distributed throughout the temperate regions of the world and are responsible for a number of economically important diseases of cultivated crops. However, they have a restricted host range and are confined to a few closely related plants in nature.

One feature of CaMV which makes it attractive as a vector is that infection becomes systemic and very high levels of the virus are found in infected cells. In order for CaMV to be transmitted through the vasculature of the plant the DNA must be encapsidated in viral protein. Thus insertion of foreign DNA into the viral genome must not interfere with virus assembly. Unfortunately the CaMV genome has virtually no non-coding regions (Fig. 13.3) in which to insert foreign DNA although two regions of the genome, genes II and VII, do not seem essential for infection. The feasibility of replacing at least one of them with foreign DNA has been demonstrated. Gene II was replaced with a bacterial gene encoding methotrexate resistance and the chimaeric molecule used to infect turnip plants. Not only did the turnips become systemically infected but they also developed resistance to methotrexate applied as a spray. No doubt other applications soon will be forthcoming.

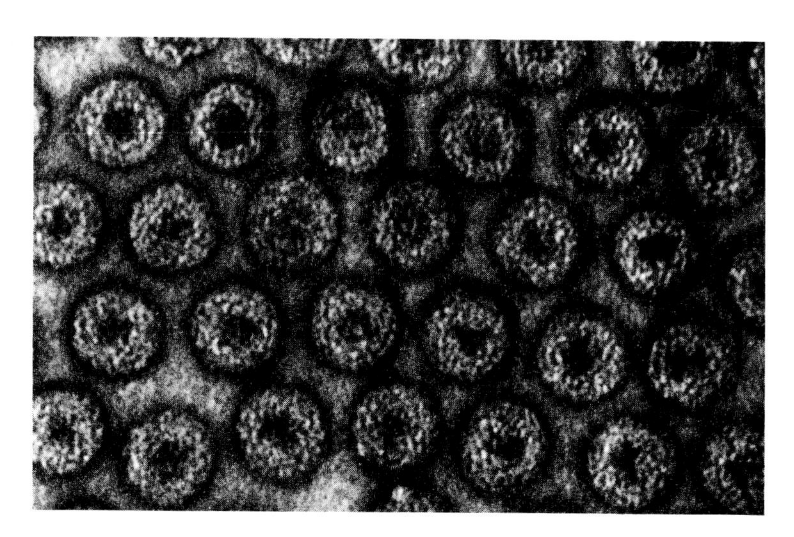

Fig. 13.2 Electronmicrograph of Cauliflower Mosaic Virus (CaMV). (Courtesy of M. Webb, National Vegetable Research Station.)

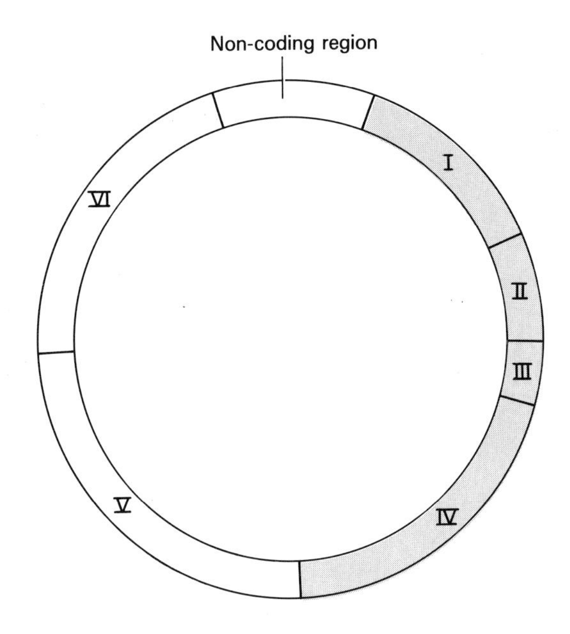

Fig. 13.3 The genetic map of CaMV. The open-reading frames are indicated by roman numerals.

Geminiviruses as vectors Geminiviruses derive their name from the fact that they have a geminate (paired particle) morphology (c.f. Gemini, the heavenly twins). They have a very small, single-stranded circular DNA genome but replication is via a double-stranded intermediate. An attractive feature of the group is that they infect a wide range of crop plants which includes both monocotyledons and dicotyledons. Since very little is known about the molecular biology of this group of viruses it is too early to speculate on their ability to be harnessed as vectors.

RNA plant viruses as vectors RNA-containing plant viruses have not been given much serious consideration as vectors because of the difficulty of joining RNA molecules *in vitro*. This problem can be overcome by making cDNA copies of the RNA prior to in-vitro manipulation. Indeed, a sequence encoding chloramphenicol-resistance (chloramphenicol acetyltransferase) has been engineered into the Brome Mosaic Virus genome in this way, although so far expression has been limited to plant protoplasts. This approach using a cDNA intermediate is slightly cumbersome but does have much to commend it. In particular, most plant viruses have genomes of RNA rather than DNA and thus the range of potential vectors is immediately expanded. Also, many of the RNA viruses have a filamentous (rod-shaped) morphology. The significance of this is that the size of the virus particle is determined by the length of the viral nucleic acid, i.e. there are no size restrictions on the nucleic acid to be packaged.

The disadvantage of viral vectors As pointed out earlier, a key requirement for plant breeding is that the introduced genes be transmitted to subsequent generations. With virus vectors this is problematical. If the virus is not seed transmitted, then the modified plants will need to be vegetatively propagated. Although this might be acceptable with certain crops, e.g. potatoes, there is always the possibility that the genetic information carried by the virus will be lost. It would be more satisfactory if the introduced genes could be introduced into a plant chromosome.

TI PLASMIDS AS VECTORS

Crown gall is a plant tumour which can be induced in a wide variety of gymnosperms and dicotyledonous angiosperms by inoculation of wound sites with the bacterium *A. tumefaciens*. Crown gall tissue represents true oncogenic transformation; callus tissue can be cultivated *in vitro* in the absence of the bacterium and yet retain its tumorous properties. These include the ability to form an overgrowth when grafted onto a healthy plant, the capacity for unlimited growth *in vitro* in the absence of plant hormones, and the synthesis of *opines*, which are unusual amino acids not found in normal tissue. The most common of these opines are octopine and nopaline (Fig. 13.4). The type of opine produced is determined by the *Agrobacterium* strain causing the infection and not by the host plant.

Investigation of the molecular biology of crown gall disease has revealed that *A. tumefaciens* has evolved a natural system for genetically engineering plant cells. In particular, the oncogenic properties of *A. tumefaciens* reside in a plasmid known as the Ti (tumour-inducing) plasmid. When the bacterium infects plant cells a portion of this DNA, called T-DNA, is transferred to the plant genome and it is the presence of the T-DNA which renders the cell oncogenic. The processes of DNA transfer and tumour formation are distinct (Fig. 13.5). The *vir* region of the Ti-plasmid encodes diffusible products which facilitate DNA transfer to the plant.

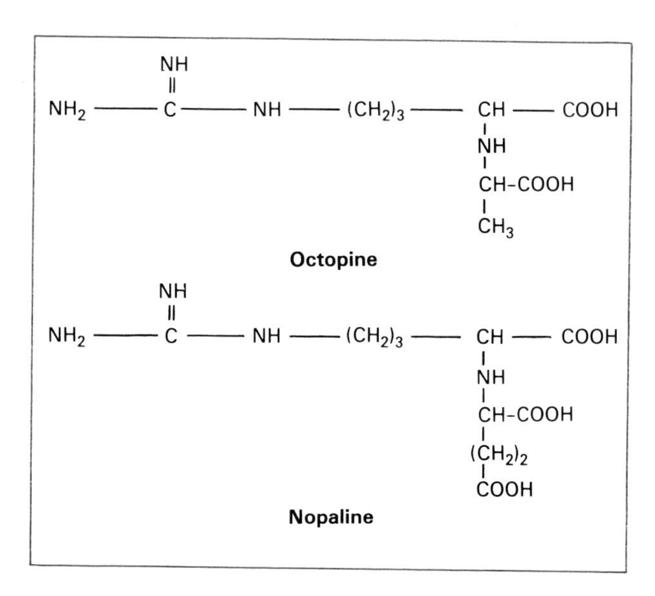

Fig. 13.4 The structure of the unusual amino acids octopine and nopaline.

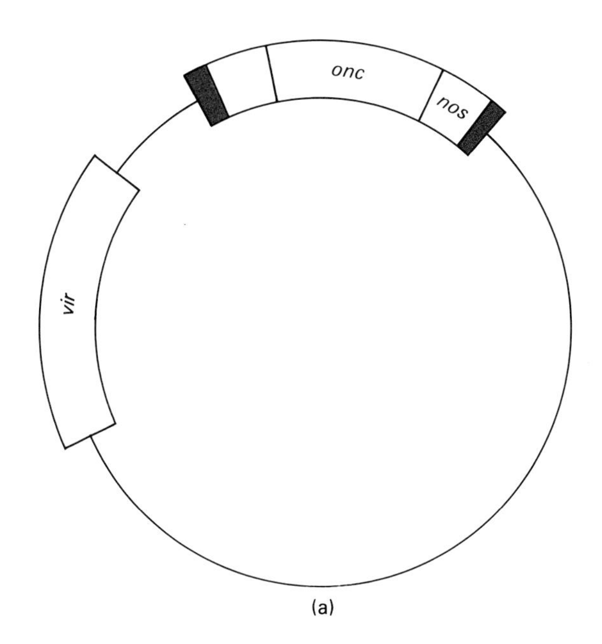

The *onc* genes of the T-DNA itself function within the plant cell to produce and maintain the oncogenic phenotype. The *onc* genes can be deleted from the T-DNA without affecting DNA transfer. However, for DNA transfer to occur the boundary sequences between the T-DNA and the rest of the Ti-plasmid must be intact.

Initial attempts to use the Ti-plasmids as plant vectors proved difficult because of the large size of the plasmids (mol. wt 200×10^6) and their genetic complexity. Consequently a system of intermediate vectors has been developed. Such vectors are bifunctional, being able to replicate in *Escherichia coli* and *A. tumefaciens* and contain all or part of the T-DNA. The intermediate vector is manipulated *in vitro*, transformed into *E. coli* and then transferred to *A. tumefaciens*. In the latter host in-vivo recombination will occur and the T-DNA of the resident Ti-plasmid will be replaced with the mutant or genetically altered T-DNA. In this way a whole variety of foreign genes, e.g. interferon, ovalbumin and leghaemoglobin, have been transferred to tobacco cells.

For a foreign gene to be expressed in plant cells it is essential that it be preceded by a promoter recognized by the host cell. Since opine synthesis is not essential, vectors have been constructed in which the foreign gene is placed under the control of the *nos* (nopaline synthase) promoter. Thus transformed cells have been obtained which express antibiotic-

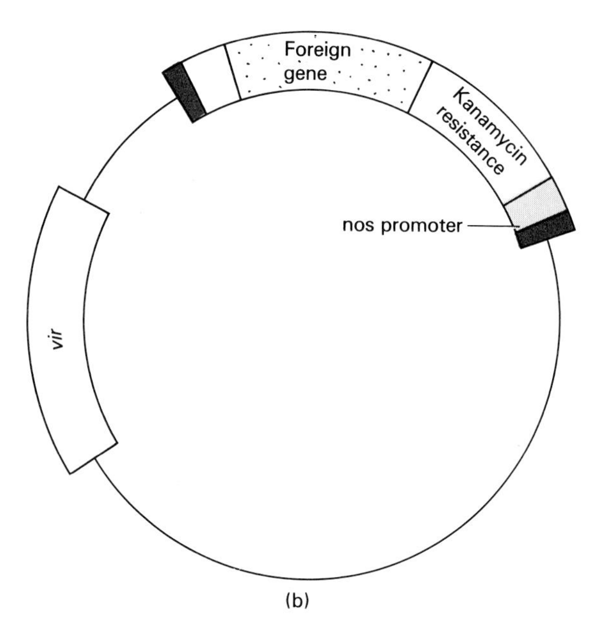

Fig. 13.5 (a) Schematic representation of a Ti-plasmid. *vir* is the virulence region whose presence is essential for T-DNA transfer to the plant genome. *nos* is the gene encoding nopaline synthase. *onc* is the group of genes whose gene products are responsible for tumour formation. The bold red areas are the DNA sequences essential for T-DNA transfer.
(b) A modified vector for plant gene manipulation. The kanamycin-resistance gene under the control of the nopaline synthase promoter is essential for selection of transformed cells.

resistance genes derived from bacteria. Ultimately the introduced genes need to be expressed in the correct tissue at the correct time. Progress is being made. When the antibiotic resistance gene was placed under the control of the promoter from the gene for the small subunit of ribulose bisphosphate carboxylase, the level of antibiotic resistance observed was regulated by light as expected.

Another requirement for plant genetic engineering is the ability to transfer the cloned gene to every cell in the plant and for it to be passed to subsequent plant generations. This is not possible if genetically engineered *A. tumefaciens* is used directly to infect whole plants; the foreign gene will be restricted to the tumour. Furthermore, regeneration of plants from tumour tissue is difficult. The solution is to delete the *onc* genes from the T-DNA. In the absence of the *onc* genes transfer of T-DNA to plant cells is detected by selection for a cloned gene, e.g. antibiotic resistance. This technique has been further improved. Instead of infecting whole plants with *A. tumefaciens* the bacteria are co-incubated with leaf discs placed on tissue-culture medium supplemented with antibiotic. Those plant cells which become infected acquire antibiotic resistance and grow out of the disc as a callus. This callus subsequently is used to regenerate plants. An alternative strategy is to include a functional alcohol dehydrogenase (ADH) gene in the T-DNA and use an ADH-negative mutant cell as host. Selection comes about by virtue of the fact that ADH-minus cells cannot grow anaerobically.

The value of the Ti-system is demonstrated by the fact that not only is the chimaeric gene present in all the cells of a plant regenerated from callus, but that such plants set seed and the chimaeric gene is transmitted in that seed. The biggest drawback to the Ti-system is that, with one exception, *Agrobacterium* does not infect monocotyledonous crop plants such as maize, rice and wheat. One way of circumventing this problem is to attempt direct gene transfer of DNA to protoplasts and then to regenerate intact plants. This approach has been used successfully. A fragment of DNA containing a kanamycin-resistance gene under the control of a CaMV promoter was introduced to tobacco protoplasts. Plants regenerated from transformed cell lines appeared normal and the kanamycin-resistance phenotype was not only expressed but segregated in a Mendelian fashion.

WHAT CAN BE ACHIEVED?

Herbicides are widely used to clear agricultural land before crops are sown but unwanted seeds which subsequently land on this ground can germinate. The resulting weeds not only compete with the crop for nutrients but can, if present in high numbers, render the crop unusable. If the crops were engineered to be herbicide-tolerant, then the farmer could spray them with herbicide as required. One widely used herbicide is glyphosate ('Tumbleweed', 'Roundup'), which inhibits EPSP synthase, an enzyme involved in aromatic amino acid biosynthesis. Two approaches have been used to engineer glyphosate-resistance in plants. In one, a bacterial gene encoding glyphosate-resistant EPSP synthase was engineered into tobacco plants. In the other, the level of expression of EPSP synthase in *Petunia* sp. was increased effectively to 'titrate out' the glyphosate. In both instances the plants exhibited increased tolerance to glyphosate when sprayed with the herbicide.

Plants supply over 70% of the protein required by man and animals but many plant proteins are deficient in essential amino acids. Thus cereals have a low lysine and threonine content whereas legumes are deficient in methionine and cysteine. Rice, on the other hand, has a reasonable amino acid balance but has a low overall protein content. These deficiencies do not matter if several plant proteins combined make up the diet, e.g. a legume plus a cereal, or if the diet is supplemented with animal protein. Unfortunately, in many parts of the world a single plant species can be the staple food and in such places malnutrition, due in part to lack of one or more essential amino acid, is common. By means of protein engineering it should be possible to alter the amino acid composition of plant storage proteins. Already the genes for a number of plant storage proteins have been cloned (Table 13.4). If suitable biophysical data on these proteins is available, it is probable that the changes made will have little effect on texture and cooking properties.

A second target for protein engineering is the enzyme ribulose bisphosphate carboxylase (RuBPCase). This enzyme mediates the key step of carbon dioxide fixation (Fig. 13.6) but it can also catalyse the addition of oxygen to ribulose bisphosphate. This latter process is known as photorespiration and, because carbon dioxide is liberated, reduces the

Table 13.4 Some useful plant proteins whose genes have been cloned.

Protein	Source	Function/use
β-conglycinin	Soybean	Seed storage protein
Glycinin	Soybean	Seed storage protein
Hordein	Barley	Seed storage protein
Legumin	Pea	Seed storage protein
Phaseolin	French bean	Seed storage protein
Vicillin	Pea	Seed storage protein
Zein	Maize	Seed storage protein
RuBPCase, large subunit	Tobacco, wheat	CO_2 fixation
RuBPCase, small subunit	Pea	CO_2 fixation
Thaumatin	*Thaumatococcus*	Sweet-tasting protein

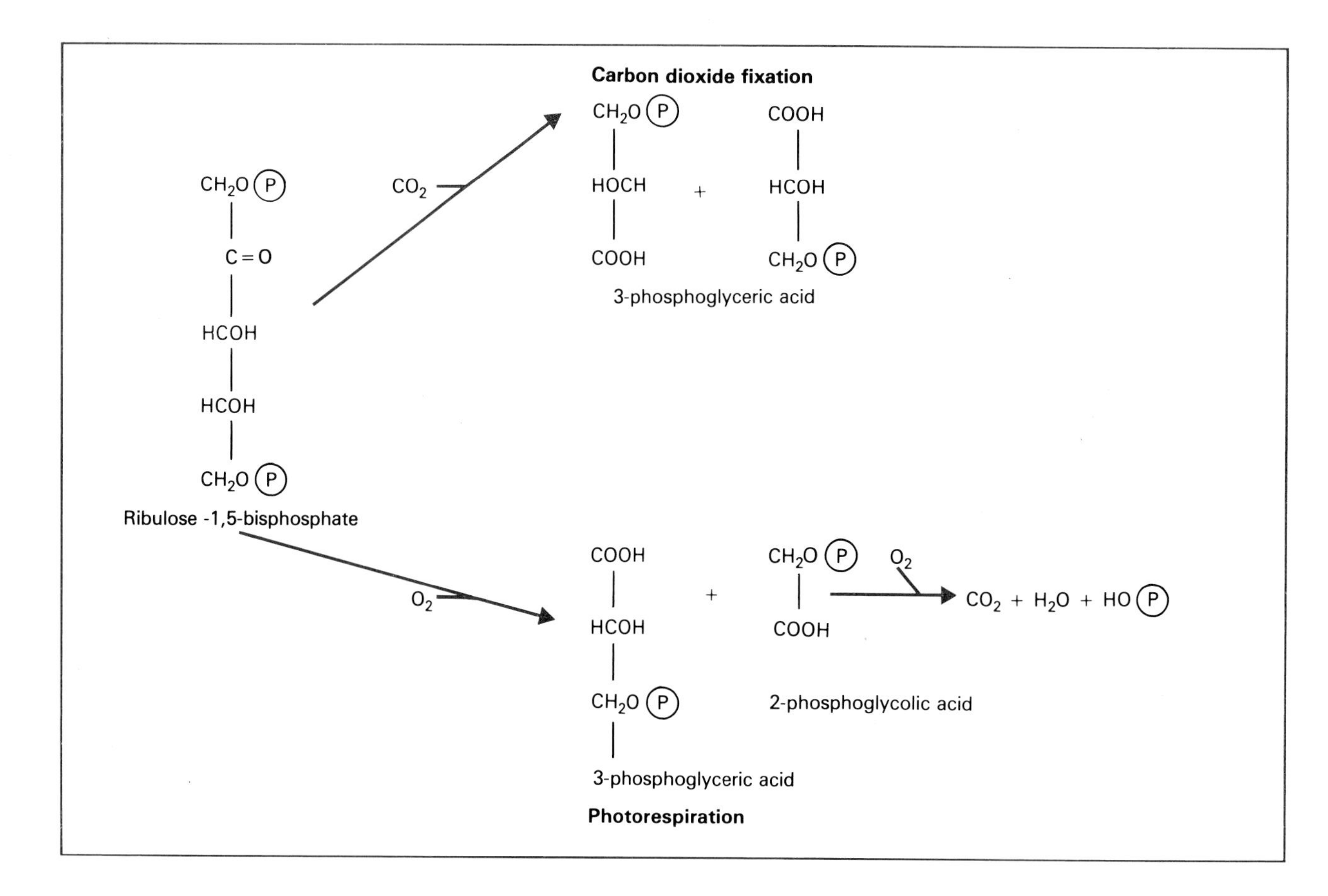

Fig. 13.6 The different products of RuBPCase action on ribulose-1,5-bisphosphate.

efficiency of photosynthesis. Although it is unlikely that it will be possible to separate the carboxylase and oxygenase functions, site-directed mutagenesis could yield enzymes with a higher affinity for carbon dioxide and a lower affinity for oxygen. In this respect they would be analogous to the modified interferons described earlier (see p.43).

Further reading

GENERAL

Evans D.A. (1983) Agricultural applications of plant protoplast fusion. *Bio/technology* **1**, 253–61.
Lichtenstein C. & Draper J. (1985) Genetic engineering of

plants. In Glover D.M. (ed.) *DNA Cloning*, Vol. II, pp. 67–120. IRL Press, London.
Old R.W. & Primrose S.B. (1985) *Principles of Gene Manipulation*, 3rd edn. Blackwell Scientific Publications, Oxford.
Rao A.S. & Singh R. (1986) Improving grain protein quality by genetic engineering: some biochemical considerations. *Trends in Biotechnology* **4**, 108–10.

SPECIFIC

Brisson N., Paszkowski J., Penswick J.R., Gronenborn B., Potrykus I. & Hohn T. (1984) Expression of a bacterial gene in plants using a viral vector. *Nature* **310**, 511–14.
Flick C.E., Kut S.A., Bravo J.E., Gleba Y.Y. & Evans D.A. (1985) Segregation of organelle traits following protoplast fusion in *Nicotiana*. *Bio/technology* **3**, 555–60.
Fraley R.T., Rogers S.G., Horsch R.B., Eichholtz D.A., Flick J.S., Fink C.L., Hoffman N.L. & Sanders P.R. (1985) The SEV system: a new disarmed Ti plasmid vector system for plant transformation. *Bio/technology* **3**, 629–35.
French R., Janda M. & Ahlquist P. (1986) Bacterial gene inserted in an engineered RNA virus: efficient expression in monocotyledonous plant cells. *Science* **231**, 1294–7.
Grimsley N., Hohn T., Davies J.W. & Hohn B. (1987) *Agrobacterium*-mediated delivery of infectious maize streak virus into maize plants. *Nature* **325**, 177–9.
Paszkowski J., Shillito R.D., Saul M., Mandak V., Hohn T., Hohn B. & Potrykus I. (1984) Direct gene transfer to plants. *EMBO Journal* **3**, 2717–22.

Part VI
Social Aspects of Biotechnology

14/Legal, Social and Ethical Aspects of Biotechnology

INTRODUCTION

Recent advances in biotechnology have created many public policy and legal issues. A number of the most relevant are discussed here: patents, deliberate release of recombinant organisms, ethical and legal aspects of gene therapy, legal implications of DNA fingerprinting, biological warfare, and the future of biotechnology in Third World countries. Scientists often find the practice of industrial law confusing; they are used to constructing theories based on the (supposed) rational examination of facts. Lawyers also examine facts but they interpret them within the framework of existing law. Unlike new theories, new laws are not created instantly and final acceptance comes only after months or years of courtroom debate. Reaching a final decision can take years. Nowhere is this clearer than in the field of patent law. Even though a patent examiner permits a patent to be issued, his decision may be challenged in the courts of law by a third party. Alternatively, a third party may deliberately infringe a patent in the belief that the patentee would lose his case if he chose to defend his patent rights in a court of law. Both situations have arisen recently regarding patents granted in respect of interferons. In one instance two major companies agreed out of court not to proceed with litigation because of the high costs involved and the uncertainties as to who would win.

Biotechnology and patent law

WHAT IS A PATENT?

Patents are rights granted in respect of inventions capable of industrial application. These rights are not automatic but require action to register them — the filing of a patent application. After filing the application is scrutinized by a patent examiner. If the examiner is convinced that the subject of the patent is both *novel* and *non-obvious*, i.e. involves an inventive step, the patent will be issued. Embodied within the novelty element is the absence of *prior disclosure*. In return for the legal protection given to the invention, the patentee must provide a written description of the invention which is sufficiently clear and effective to enable a person of *ordinary skill in the art* concerned to practise the invention from its teaching, i.e. a patent must be *enabling*. As normally constructed, the descriptive matter of a patent would be presented under a series of typical headings like those shown in Table 14.1.

CATEGORIES OF PATENTABLE INVENTION

Three categories of patent can be recognized and examples of each can be found in the field of biotechnology. These categories are:

1 compositions of matter (products);
2 processes for manufacture;
3 methods of use.

Product patents are the most effective form of pro-

Table 14.1 Typical descriptive sections found in a patent.

Heading	Content
Field of the invention	The technological field and the broad nature of the invention
Background of the invention	The problem to be solved, and a summary of prior art in the field and the disadvantages thereof
Objects of the invention	The nature of the improvements that the invention seeks to provide
Summary of the invention	A distillation of the essential elements of the invention
Detailed description of the invention	A chronological sequence of experimental steps to enable someone skilled in the art to repeat the work and recreate the invention
Detailed claims	A list of the specific aspects of the invention which are to be protected

tection because proof that the claim is being infringed by a competitor is usually apparent from an inspection or an analysis of the competitor's product, i.e. there is no need to inquire into the method of manufacture. In addition, this is the only category of patent which has recognition throughout the US, Japan and Western Europe. A new antibiotic normally would be covered by a product patent and it would be irrelevant whether this antibiotic were discovered by a standard screening procedure or by using gene manipulation to generate novel combinations of antibiotic biosynthetic genes (see p. 74). A second example would be a novel micro-organism that could degrade a previously recalcitrant molecule, e.g. dioxin or Agent Orange, and which could be used to clean up toxic waste spills. It should be noted that until 1980 it was not possible to patent a living organism other than a plant. However, in that year the US Supreme Court ruled (*Diamond* v. *Chakrabarty*) that the fundamental distinction between living organisms and inert matter was irrelevant for patent purposes.

Many of the products we expect to produce by recombinant DNA technology are not new, rather they are natural products, such as hormones, produced in quantity by an inexpensive procedure. As those products are not new it might be thought that they are not patentable but this is not necessarily so, as the example of US patent 2 563 794 shows. This patent has a single claim to highly purified vitamin B_{12} of defined chemical characteristics. The defendant in an infringement suit held that the patentees had not discovered vitamin B_{12} but merely recovered it from existing material in 'more nearly pure' condition. As such it was not a newly created compound but only a purified form of the anti-pernicious anaemia factor known to be present in liver. This argument was dismissed by the court on the grounds that until the patent was filed nobody had produced a product of comparable purity.

In recent years, most developed countries have adopted the product claim as an appropriate form of protection for new substances. Before this, in most countries of the world it was necessary even in the case of a new product, to claim the invention in process terms. This unsatisfactory situation, which still exists in Latin America and Eastern Europe, requires alternative methods of preparation to be devised and the use of multiple process claims to cover every practical possible pathway to the product. The genuine process invention is one where a truly novel route or method has been devised or where new conditions have been found for operating a process of generally known type in order to achieve an improved result; an example here would be the production of indigo in *Escherichia coli* strains carrying a cloned naphthalene dioxygenase gene (see p. 78). Incidentally, this is a perfect example of a discovery being non-obvious. Another example would be the improved substrate utilization by strains of *Methylophilus methylotrophus* carrying a cloned *gdh* gene (see p. 67) and which are grown for single-cell protein production.

Process patents are weak patents in that it is difficult to know when a competitor has infringed your patent. For this reason many companies choose not to patent new processes but to retain them as *trade secrets*. The danger here is that a competitor may discover the same process and patent it and could then sue you for infringement! In such a situation the legal outcome is not guaranteed. Some of the factors to be considered before making a decision regarding filing a patent or keeping a trade secret are set out in Fig. 14.1.

Where a substance is not new but has been newly

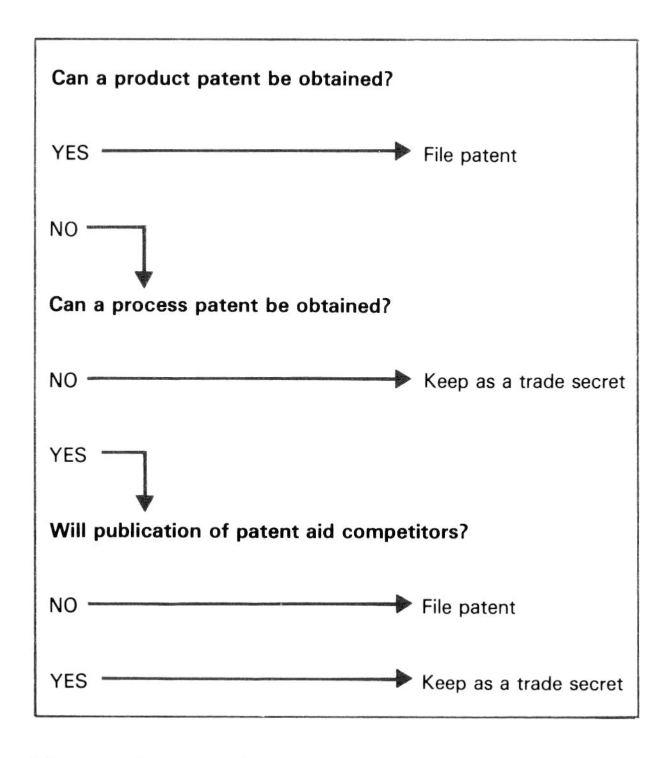

Fig. 14.1 Factors influencing the decision to patent or not to patent a product or process.

discovered to possess a valuable property it sometimes can be protected by a method of use patent. Very often this covers the substance in admixture with other substances in a form in which it will be utilized to take advantage of the new property. Patents covering the formulation of therapeutic proteins produced by recombinant DNA technology now are being published, the usual claim being increased stability.

THE UNCERTAINTY OF PATENT LAW

Patent litigation is an inexact science and infringement suits in biotechnology will often be reduced to debates between opposing expert witnesses. The situation is exacerbated by variations in patent law between different countries. For example, the requirement for absence of prior disclosure in the US means that the subject matter of the patent application should not have been described in public more than one year before the submission of the patent application. In Europe and the rest of the world patent applications must be filed before any publication. A better example, perhaps, concerns the *doctrine of equivalence*. Thus, a claim to the production of indigo in *E. coli* would not literally be infringed by an analogous process in another genus, e.g. *Pseudomonas*. However, an appeal against infringement might be upheld under the doctrine of equivalence. The courts generally allow a broader scope and wider range of equivalents to patents that represent a new field or a distinct step in the progress of an art. A countervailing doctrine, however, recognizes that the behaviour of chemical or biochemical compounds is not always predictable as conditions vary, and this should limit the scope of claims to only those which a patent specifically teaches. Because of these uncertainties it is essential to employ a skilled legal draughtsman in the preparation of a patent; ideally he should be trained in a biological science and law.

Legal implications of DNA fingerprinting

Forensic biology is the testing of biological materials, the results of which may be used as evidence in legal and criminal proceedings associated with such diverse cases as paternity disputes, sexual assault and murder. In the last two decades there have been considerable technological advances as a result of the discovery of many new genetic polymorphisms among human proteins. The use of electrophoretic and immunological techniques to identify these new polymorphisms has worked its way into the forensic science repertoire with the result that more definitive genetic typing information can be provided for the legal system. However, these methods are not ideal. In the case of sexual assault the presence of spermatozoa indicates that sexual activity has occurred. The analysis of proteins from these spermatozoa, recovered intravaginally or as a dried strain, can be used to increase the probability of making a correct identification of the assailant but a thorough analysis is difficult. Only the ABO blood group and a select number of human leucocyte antigens (HLA) have been typed on sperm and the polymorphic red cell enzymes found in semen are limited in their phenotypic diversity. As a result, the likelihood of excluding a falsely accused individual is low; alternatively, there is a high probability that a random individual from the population might have the same markers as the assailant. Against this background what is the forensic potential for the DNA polymorphisms which can be detected with gene probes (see p. 32)?

The scientific conditions which must be satisfied before any blood group or other protein polymorphism can be used in a forensic science context are listed in Table 14.2. Since these protein polymorphisms have a genetic basis it is logical that DNA polymorphisms satisfy the same criteria. In the case of simple restriction fragment length polymorphisms (RFLP), like those shown in Fig. 3.7 for sickle-cell anaemia, there is no problem in fulfilling conditions 1–3. However, they are subject to the same criticisms as protein polymorphisms; that is, their lack of phenotypic diversity means that their diagnostic potential is limited. The diagnostic potential of the much more complex RFLP exhibited by minisatellite DNA (Fig. 3.10) is far greater but it is much harder to satisfy the criteria laid down in Table 14.2. The available data strongly suggest the true genetic character of these polymorphisms but absolute proof will be not available for some time. However, there is no way that gene frequencies for the major population groups can be established since the profile for each individual is expected to be unique. The uniqueness of these DNA 'fingerprints' can be truly established only by testing all individuals, both living and dead. Clearly this is not possible but then neither is it for conventional

Table 14.2 Criteria for the use of any protein polymorphism in a forensic science context. (Adapted from Sensabaugh (1986)).

1 The marker used must be validated as a true genetic character. Its mode of inheritance must be demonstrated by family studies and it must be shown to be stable in an individual over a lifetime
2 Gene frequencies for the major population groups must be established to provide a base for the interpretation of findings
3 There must be a nomenclature to define the variants in each polymorphic system; this is necessary both for purposes of record keeping and for communication of data
4 Standard types and typing reagents must be available to the forensic science community to allow independent testing in different laboratories
5 Methodological guidelines need to be defined to ensure reliability of test results
6 Ultimately, markers must pass the barrier of blind trial testing. This establishes that the marker does not possess inherent ambiguities that might lead to typing error. The nature of the blind trials depends on whether the targeted use is paternity testing or criminal evidence

fingerprints. Useful information may come from the US military. They are considering having all their personnel DNA fingerprinted to make it possible to identify from tissue fragments the victims of air disasters or bombings. Clearly this will require a system of nomenclature and one possibility is to scan autoradiographs in a densitometer and classify the patterns so produced in the same way as infrared or mass spectra.

Although the techniques involved in DNA fingerprinting are routine in molecular genetics laboratories, they are much too complex for any routine diagnostic laboratory. A British chemical and pharmaceutical company, Imperial Chemical Industries (ICI), is setting up a number of test centres in Europe and the US which will provide the necessary expertise. There is a forensic problem associated with this because the prosecution and defence should have access to independent tests. There is considerable unease about there being only one centre to which samples can be sent. One solution in Britain would be for the Home Office Forensic Science Service to carry out 'fingerprinting' for the police, while the defence would use the ICI centre. Alternatively, separate scientists acting for the two sides could have independent access to the same facility as sometimes occurs already in forensic science laboratories.

Deliberate release of genetically engineered organisms

GENETICALLY ENGINEERED MICROBES

The very first commercial applications of recombinant DNA research have used genetically engineered microorganisms in fermentation technologies to make useful pharmaceutical products. Soon a host of other commercially useful products derived from gene manipulation will reach the marketplace. While there has been some debate about the large-scale culture of recombinant organisms, particularly with regard to the safety of workers and the consequences of accidental release, concerns are abating rapidly. Partly this is due to the fact that modern fermentation plant can be made virtually leak-proof and the organisms killed at the end of the process; partly it stems from the fact that recombinant strains of *E. coli*, the current organisms of choice, do not have a propensity to colonize the gut. By exercising relatively simple precautions, companies can ensure that the risks posed by fermentation of well-characterized recombinant strains are not appreciably greater than the risks associated with traditional fermentations.

Genetically engineered microorganisms can also be commercially useful in ways that require that they be introduced on a large scale into the natural environment, i.e. *deliberate release*. Applications include pollution control, tertiary oil recovery and, most controversial of all, ice nucleation-defective bacteria for frost damage control (see Box). The feature which causes most concern is that the microorganisms may proliferate in the environment. Indeed, proliferation is often part of the *modus operandi* of the application. If, after release, we discover some unanticipated detrimental effects, there is no way we can destroy all the offending microbes. Not only can we not see them, we may have difficulty in monitoring their presence although this might be facilitated by the use of suitable genetic markers. Another feature which needs to be considered is that a technology that is beneficial to one person can be detrimental to another, as attempts at weather modification by cloud seeding have shown.

Bacterial ice nucleation

Frost-sensitive plants cannot tolerate ice crystal formation within their tissues. Ice crystals within sensitive plant tissues propagate rapidly, both intercellularly and intracellularly, causing mechanical breakdown of plant tissue and subsequent death. Many liquids, including water, do not invariably freeze at the melting point of the solid phase. These liquids can be supercooled below the melting point of the solid phase, e.g. water can be supercooled to −10 °C to −20 °C. The water–ice phase transition requires the presence of a catalyst or *ice nucleus*. Plants do not have intrinsic ice nuclei active at temperatures above −5 °C but certain bacterial species can act as ice nuclei and thus have a primary role in limiting supercooling and inciting frost damage to frost-sensitive plants. The commonest ice nucleation-active bacteria isolated from plants are *Pseudomonas syringae* and *Erwinia herbicola*. They can initiate ice formation at temperatures of −1.5 °C to −5 °C, most probably by means of an outer membrane component. The gene(s) for ice nucleation have been cloned and are under active study.

Ice nucleation-active bacteria are present in large numbers in all temperate regions of the world and may be important in initiating rain and snow. Currently these bacteria are used to facilitate the formation of 'artificial' snow on ski slopes. Water containing *Ps. syringae* is sprayed through a fine nozzle onto a fan and the expansion-induced cooling produces snow.

An aspect of deliberate release which seems to be forgotten is that for over 50 years microbiologists have been adding cultures of microorganisms (*inoculants*) to the environment in the hope of promoting a beneficial change, e.g. nodulation of legumes, oil and chemical waste removal, plant residue decomposition and destruction of plant pests. No substantial damage of any significance has been caused through such practices. There is no reason to think that a bacterium or fungus that is known not to damage the environment will cause environmental problems after it has obtained several well-characterized foreign genes. Although genetically engineered microbes may exchange their genetic information with indigenous soil organisms this should be of little concern; unrelated microbes have been sharing and reassorting their genes since time immemorial using the same methods as geneticists, e.g. transposon mutagenesis, phage-mediated transduction and plasmid transformation and conjugation. There are many plant pathogens that can naturally exchange genes with *E. coli* but we have yet to isolate a phytopathogenic *E. coli*.

Another question which must be asked concerns the ability of a newly introduced microorganism to predominate in a particular habitat already filled with a complexity of microbes which have evolved to compete successfully for all available niches. Concerns about the competitiveness of deliberately released microorganisms stem from experience with the introduction of exotic weed species whose proliferation occasionally has been uncontrollable. However, in an area of natural habitat of, e.g. an acre, there may be 10–50 different plant species; there could be many more different genera of microorganisms on a single plant. Regardless of whether the environmental dangers associated with deliberate release are real or imaginary it makes good sense to follow testing protocols before an organism is released. But how does one design relevant tests when environmental damage can only be assessed in field tests? For this reason there is a demand for an '*environmental impact*' statement before approval for deliberate release can be given.

Although environmental risk assessment is believed to be analogous to the human health risk assessment currently practiced by institutional biosafety committees, nothing could be further from the truth: man essentially is a single habitat, the environment is not. The requirements for an environmental impact statement prior to deliberate release of a recombinant leads to bizarre situations. For example, the current controversy surrounding deliberate release stems from the wish of an investigator to attempt to stop frost damage to potato plants by spraying them with a genetically engineered strain of *Ps. syringae*. The strain in question is recombinant only in so far as gene manipulation was used to introduce a *deletion* in a gene for ice nucleation; this deletion renders the strain non-phytopathogenic. Whereas deliberate release of this organism has been subject to complex litigation, there is nothing to stop the investigator from deliberately releasing a mutant with similar properties created by conventional microbial genetics!

The US Environmental Protection Agency (EPA) has drawn up a document on 'Proposed Points to Consider for Environmental Testing of Microorganisms'. Figure 14.2 shows the likely approval process.

GENETICALLY ENGINEERED PLANTS AND ANIMALS

There are fewer problems associated with the 'deliberate release' of genetically engineered plants and animals. For a start, since only large (crop) plants and domestic animals will be subject to gene manipulation, their spread in the environment can be observed. This contributes significantly to peace of mind. Since farm animals normally are kept in captivity, transgenic cows or pigs generate little concern. Transgenic fish could raise problems because the introduction of certain fish species as biological control agents has occasionally had disastrous consequences.

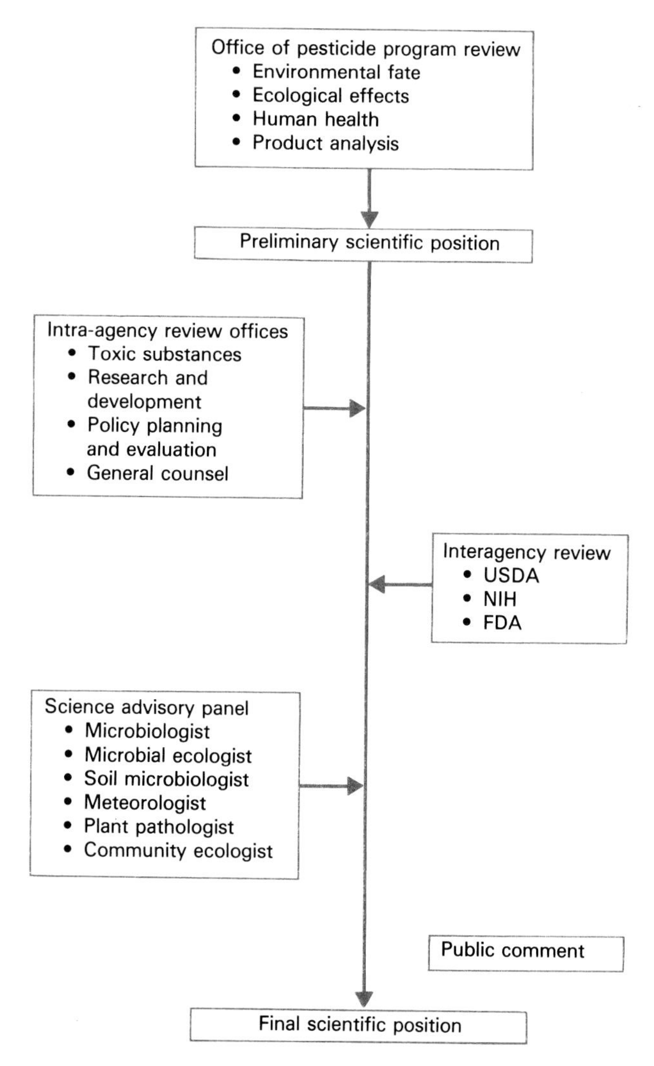

Fig. 14.2 Review procedure to be used in the US by the Environmental Protection Agency before agreeing to the deliberate release of recombinant microorganisms.

Some agronomists claim that the introduction of genetically engineered plants is analogous to the importation of exotic plant species and could disturb the existing ecological balance in an unpredictable way. But are not the major crops of the Western World, e.g. soybeans, wheat and rice in the US, foreign introductions and has their cultivation not been environmentally harmless? The consequences inherent in the release of genetically engineered crop varieties appear to differ little from those following the release of any conventionally bred new variety. There appear to be two concerns. The first arises because many commercial plant varieties are improved by crossing them with related weedy species. Therefore genetic engineering efforts to accomplish herbicide tolerance and increase photosynthetic efficiency could, as a result of introgression, result in fitter weeds. Such concerns seem misplaced if plant breeders are using the traditional methods to achieve the same ends. The second issue arises because the vectors used for genetic engineering of plants are derivatives of pathogenic microorganisms. If experiments are done with vectors which are not autonomous replicons, and if the newly introduced DNA is stably integrated, genetically engineered plants should not serve as sources for disease transmission in the field.

RECOMBINANT ORGANISMS AND THE MANUFACTURE OF FOOD AND DRUGS

In the next few years it is anticipated that a whole series of new drugs, principally human proteins and viral vaccines, will be available to the physician as a result of developments in recombinant DNA technology. Should these drugs be regulated any differently to existing drugs produced by more conventional means? Within the drug regulatory agencies there is a growing recognition that this should not be the case. The reason is simple: what is important is the quality of the drug, not its route of manufacture. Of course, part of the quality specification must include freedom from infectious agents and this includes nucleic acid.

With drugs applied to domestic animals the situation is a little different. If the animal is to be used ultimately as a source of human food, then the drug must not contaminate the slaughtered animal. This poses an interesting question in the case of animal hormones. To what extent are low levels of hormone detected at slaughter due to endogenous synthesis rather than administration? The answer may well depend on the sensitivity of the test employed; if no hormone is detected there is no problem!

A potential regulatory issue which has received little public debate is the use of recombinant micro-organisms to produce food. For example, to facilitate cheese manufacture strains of lactic acid bacteria which are resistant to antibiotic residues in milk and to phage attack, or which stably maintain their protease and phosphogalactosidase genes (see p. 65) could soon be available. Will the use of such strains be acceptable to regulatory agencies? Also, fungal proteases have long been used as substitutes for calf rennet (chymosin); now that chymosin produced in *E. coli* has been found to produce gastronomically acceptable cheese will its commercial use be approved?

FINAL COMMENT

Given the considerations outlined above, how does one assess the current European controversy surrounding a supposedly safe vaccine against rabies? The new vaccine is a version of the vaccinia virus, which has long been used to prevent smallpox, but which has been engineered to express a protein from rabies virus. Target populations are foxes in Europe and raccoons on the east coast of the US. If these animals were to be vaccinated, there would be no problem(?) but the idea is to feed the vaccine in bait!

Gene therapy

At the time of writing, no experiments on *human* gene therapy are under way, but the first of these is expected within the next 1–2 years. Almost certainly these experiments will take place in the US and the National Institute of Health (NIH) is gearing up its administrative apparatus in preparation for receipt of the first protocols. These protocols will have to be approved by three groups — the investigators local institutional review board and institutional biosafety committee, the NIH Working Group on Gene Therapy, and the full Recombinant DNA Advisory Committee — prior to submission to the directors of the NIH for final review. At all federal stages the review process will take place in the open. Not only will NIH committee meetings be open, but a precis of the protocol itself will be published in the Federal Register for public comment.

In early 1985 the NIH Working Group on Gene Therapy published draft guidelines in a 'Points to consider' document relating to the preparation of experimental protocols for human studies. On the basis of public comment this draft has been revised but what has been interesting is the *lack* of public comment. One interesting point did emerge. The guidelines apply explicitly to 'recombinant DNA and DNA derived therefrom'. They do not apply to recombinant RNA which, technically, is outside the charter of the Recombinant DNA Advisory Committee. Since RNA viruses (retroviruses) are likely to be the vectors used to transfer genetic material into the cells being returned to the human body this presents an interesting legal situation!

Three aspects of the modified 'Points to consider' document deserve comment. First, only experiments involving somatic cell therapy will be considered. The group will not consider germ line therapy protocols until somatic cell therapy has progressed and public discussion of the implications of germ line work has been broadened. Second, the patients will have to agree to at least a 3–5 year follow-up and to an autopsy in the event of death. Third, although investigators will be asked for results of experiments in non-human primates, results from experiments in other small mammals, e.g. mice or dogs, may be considered sufficient. There is no *prima facie* evidence that primate studies will provide information that cannot be obtained from other species.

What are the ethical considerations relating to gene therapy? Essentially, all observers have stated that somatic cell gene therapy for a patient suffering from a serious genetic disorder would be ethically acceptable if carried out under the same strict criteria that cover other new and experimental medical procedures. There are three general requirements. It will be essential to show in animals that:

1 the new gene can be put into the correct target cells and will remain there long enough to be effective;

2 the new gene will be expressed in the cells at an appropriate level;

3 the new gene will not harm the cell or the intact animal.

These criteria are similar to those required prior to the use of any new drug: the treatment should get to the area of disease, correct it, and do more good than harm. Some flexibility may be required for some disease states, e.g. the young child with ADA deficiency for which the prognosis is certain death. Although it would be unrealistic to expect a complete cure from initial attempts at gene therapy, there must be good animal data which indicate that some amelioration of the biochemical defect is likely. Even then it will be necessary to weigh the potential risks to the patient, including the possibility of producing a pathologic virus or a malignancy, against the anticipated benefits to be gained from a functional gene. For each patient there will need to be a risk-to-benefit determination but this is a standard procedure for all clinical research protocols.

A variant of somatic cell gene therapy is enhancement genetic engineering. This is not therapy of a genetic disorder but the insertion of an additional gene which produces change in some characteristic that the recipient individual desires. An example would be an additional gene for human growth hormone to produce the large size favoured by US basketball and football players. Ethically, this should not be done. Not only is there no medical need, but the short- or long-term consequences for the individual cannot be predicted. The risks are great and the real benefit nil.

The ethical and legal issues surrounding somatic cell gene therapy are really no different from those relating to heart transplant surgery. This is not the case with germ line therapy and before such therapy can be considered we will need to determine if the transmitted gene itself, or any side-effects caused by its presence, adversely affect the immediate offspring or their descendants. Satisfactory evidence that there are no long-term effects will require animal studies involving the study of several generations of progeny. The critical ethical question is whether we should ever provide a treatment which produces an inherited change and could perpetuate in future generations any mistake or unanticipated problems.

Three criteria will need to be satisfied to justify germ cell gene therapy. First, there should be considerable previous experience with somatic cell gene therapy which clearly establishes its effectiveness and safety. Second there should be adequate animal studies that establish the reproducibility, reliability and safety of germ cell gene therapy using the same vectors and procedures that would be used in humans. It should be borne in mind that, as currently practised with transgenic animals, gene therapy does not remove or destroy the defective gene in the recipient, rather, it adds a normal gene to the total DNA complement. Thus, it will be essential to know that the new DNA can be inserted exactly as predicted, that it does not destroy the functioning of another essential gene, and that it is expressed in the correct tissues at the correct time. Finally, since the societal gene pool will be affected by germ cell gene therapy the public will need to have a thorough understanding of this form of treatment and give its approval. Ethical issues aside, germ cell gene therapy could raise legal issues more complex than those currently surrounding the long-term teratological effects of exposure to nuclear tests in the South Pacific and the defoliant Agent Orange in Vietnam.

Whenever gene therapy is discussed publicly there is a tendency for the popular press to raise the spectre of the creation of 'superhumans'. Fortunately, such traits as intelligence and fertility are polygenic and their manipulation by genetic engineering will not be possible for a long time. Ethically, most of us find the prospect of eugenic genetic engineering abhorrent and the reason we do so has been succinctly expressed by W.F. Anderson: 'Our knowledge of how the human body works is still elementary. Our understanding of how the mind, both conscious and subconscious, functions is even more rudimentary. The genetic basis for instinctual behaviour is largely unknown. Our disagreements about what constitutes "human-hood" are notorious. And our insight into what, and to what extent, genetic components might play a role in what we comprehend as our "spiritual" side is almost non-existent. We simply should not meddle in areas where we are so ignorant.'

Biological warfare

The construction of harmful biological agents, deliberately or otherwise, has always been acknowledged as the most extreme hazard associated with recombinant DNA technology. The evidence to date suggests that no new infectious agent has been created accidentally and it is difficult to perceive of anything more virulent than the naturally occurring Marburg agent or Lassa fever virus. Nevertheless, the possibility of a military use for gene manipulation has been of concern to many. It must be realized that, however created, infective agents are not good warfare weapons since the results which will be obtained with them cannot be predicted with any degree of certainty. Morbidity and mortality from communicable disease agents cannot be accurately forecast because of differences in population groups with respect to physiological, genetic and socio-cultural variables, nutritional status, previous exposure to infectious agents, immunization histories and various other factors. The senior military men amongst the Superpowers are well aware of these limitations. Use of biological weapons is more likely to be sanctioned by a Third World country where the prospects for the creation of a practically useful recombinant biological weapon are unlikely.

Despite the suggested military futility of biological weapons, does research into their creation still continue? Approximately half the nations of the world are signatories to a treaty resulting from a Biological Weapons Convention of 1972. As a signatory, a nation pledges never to produce 'microbial or other biological agents, or toxins, whatever their method of production'. However, the treaty does not prohibit studies on defences against chemical and biological weapons. This, in turn, does not prohibit the creation of 'new' organisms by recombinant DNA technology so that preventive measures can be developed. Given the current state of biotechnology, as outlined in this book, it would not be difficult to turn this research round in only a few days so that biological weapons were being produced. So is the spirit of the Convention being breached?

Biotechnology and the Third World

Many political leaders of developing nations believe that great personal prestige accrues to them if their government, or better still, a foreign company or government finances a high-technology venture in their country. The fact that the country might not need such a facility, nor have the infrastructure to maintain it, is irrelevant. Modern biotechnology has not escaped these political fancies. This is unfortunate. Biotechnology has much to offer these countries but it is the kind of biotechnology which does not have much glamour attached to it. Around the world some 15 million young children die annually from the synergistic effects of malnutrition and infectious disease. A very large proportion of child deaths occur from diseases for which there are suitable vaccines. Two and a half million deaths are attributable to just three diseases: tetanus, measles and whooping cough. There is some cause for hope, for in the last two years there have been major campaigns in a number of countries including Colombia, Turkey, Bolivia, Haiti and El Salvador. In the latter country the two sides in the ongoing civil war temporarily suspended hostilities to enable mass immunization of children to take place! Diarrhoeal diseases are a major cause of death and the aetiology frequently is not known making vaccine selection more difficult. However, the availability of clean water would reduce the incidence of the disease and would permit such simple remedies as oral rehydration therapy. Biotechnology can facilitate the supply of clean water but the methodology is not new.

The above arguments aside, are there any aspects of the new biotechnology which have particular relevance for developing countries? The answer is yes. In many of these countries the major part of the diet is derived from plants. Indeed, often a single crop is the staple foodstuff and this results in malnutrition since the plant storage proteins almost invariably are deficient in one or more amino acids. By combining protein engineering techniques (see Chapter 4) with modern plant breeding methods it should be possible to develop plant varieties with enhanced levels of more nutritious storage protein as well as ones with enhanced disease resistance. The real attraction is that these plants can be genetically engineered in a laboratory anywhere in the world. Once created they can be moved to a low-technology environment for they can be propagated by unskilled workers using traditional agricultural methods.

Final comment

When the techniques of gene manipulation were first developed just over a decade ago, I, like many others, anticipated a bright future for a modern biotechnology industry. However, at the time of writing few companies have made much money from recombinant-derived products. Admittedly, there has been more success with monoclonal antibodies. Clearly the situation will improve in the next decade but as I review the contents of this chapter I wonder if the people really to benefit from the biotechnology boom will not be those in the legal profession. The lawyers amongst my friends see nothing wrong in this!

Further reading

GENERAL

Adler R.G. (1984) Biotechnology as an intellectual property. *Science* **224**, 357–63.

Benson R.H. (1986) Biotechnology patent pitfalls. *Bio/technology* **4**, 118–20.

Benson R.H. (1986) Patent wars. *Bio/technology* **4**, 1064–70.

Goodman R.M. (1985) Bringing new technology to old world agriculture. *Bio/technology* **3**, 708–9.

Hirano S.S. & Upper C.D. (1985) Ecology and physiology of *Pseudomonas syringae*. *Bio/technology* **3**, 1073–8.

Lindow S.E. (1983) The role of bacterial ice nucleation in frost injury to plants. *Annual Review of Phytopathology* **21**, 363–84.

SPECIFIC

Anderson R.M. (1986) Rabies control: vaccination of wildlife reservoirs. *Nature* **322**, 304–5.

Anderson W.F. (1985) Human gene therapy: scientific and ethical considerations. *Journal of Medicine and Philosophy* **10**, 275–91.

Brill W.J. (1985) Safety concerns and genetic engineering in agriculture. *Science* **227**, 381–4.

Gibbs J.N. (1986) Regulating bioengineered veterinary drugs. *Bio/technology* **4**, 414–16.

Hauptli H., Newell N. & Goodman R.M. (1985) Genetically engineered plants: environmental issues. *Bio/technology* **3**, 437–42.

Levin M.A., Seidler R., Borquin A.W., Fowle J.R. & Barkay T. (1987) EPA developing methods to assess environmental release. *Bio/technology* **5**, 38–45.

Sensabaugh G.F. (1986) Forensic biology — is recombinant DNA technology in its future? *Journal of Forensic Science* **31**, 393–6.

Index